POLYESTER - PRODUCTION, CHARACTERIZATION AND INNOVATIVE APPLICATIONS

Edited by **Nurhan Onar Camlibel**

Polyester - Production, Characterization and Innovative Applications
http://dx.doi.org/10.5772/intechopen.69941
Edited by Nurhan Onar Camlibel

Contributors

Hatice Kubra Kaynak, Esin Sarıoğlu, Canan Saricam, Nazan Okur, Muhammad Azeem Munawar, Shahzad Maqsood Khan, Dirk Wolfram Schubert, Nafisa Gull, Atif Islam, Muhammad Atiq Ur Rehman, Monika M Voigt, Ewa Janina Kicko-Walczak, Grazyna Rymarz, Ahmad Nawaz, Bilal Islam, Muhammad Sadiq Khattak, Weiguo Mao, Nurhan Onar Camlibel

Published by InTech
Janeza Trdine 9, 51000 Rijeka, Croatia

Publishing Process Manager Marina Dusevic
Technical Editor SPi Global
Cover InTech Design team

First published March, 2018
Printed in Croatia
Legal deposit, Croatia: National and University Library in Zagreb

Additional hard copies can be obtained from orders@intechopen.com

Polyester - Production, Characterization and Innovative Applications, Edited by Nurhan Onar Camlibel
p. cm.
Print ISBN 978-953-51-3881-5
Online ISBN 978-953-51-3882-2

For EU product safety concerns:
IN TECH d.o.o., Prolaz Marije Krucifikse Kozulić 3, 51000 Rijeka, Croatia,
info@intechopen.com or visit our website at intechopen.com.

Meet the editor

Dr. Nurhan Onar Çamlibel is an Associate Professor and has a BSc degree (2000) in Textile Engineering from the Ege University and MSc (2003) and PhD (2009) degrees in Textile Engineering from the Dokuz Eylul University. She is employed at the Department of Textile Engineering, Engineering Faculty, Pamukkale University since 2001. She received the title of Associate Professor at the Department of Textile Engineering, Engineering Faculty, Pamukkale University, in 2018. She teaches graduate and undergraduate-level courses in textile chemistry; chemical textile testing; nanotechnological applications in textile, coating, and lamination; textile degradation; and analysis methods. Her main research interests include textile treatments, functional finishing processes, enzymatic applications in textile, sol-gel applications in textile, conductive textiles, fire-protective textiles, and electromagnetic shielding efficiency of textiles. She has authored over 35 scientific publications and 1 book chapter in the aforementioned areas and coauthored over 50 conference papers and has been a manager and a researcher in national and international projects. She is a coordinator of Erasmus mobility program for students.

Contents

Preface

The preface deals with the subject of the book "Polyester: Production, Characterization, and Innovative Applications." The book consists of six chapters covering recent developments in polyester recycling and unsaturated polyester resins and composites, particle-filled polyester composites, use of polyester in construction and automotive applications, and thermal stability and fire-retardant properties of polyester.

Chapter 1 is the introductory chapter, which presents general information about polyester polymer, its production and categorization, and the properties of thermoset and thermoplastic polyester. Furthermore, the chapter emphasizes the importance of polymer for some technical applications and touches on the significance of the recycling process of polyester.

Chapter 2 mentioned the importance of polyethylene terephthalate (PET) recycling and emphasized the use of recycled PET in the textile fiber production. Researchers produced PET/cotton blend yarns with ring and compact spinning systems from PET bottle waste recycled by mechanical recycling method. Moreover, virgin PET/cotton blend fibers were produced to compare, and unevenness, imperfection index, hairiness, and tensile properties of these fibers were evaluated. The effect of raw materials, yarn spinning methods, and blend ratio on the mechanical properties of these fibers was statistically analyzed using multivariate analysis of variance. It was also stated that quality issues of textile fibers produced from recycled fibers and difficulties in filament yarn production from these fibers will be overcome in due course.

Chapter 3 indicated that polymer composites prepared from unsaturated polyester resin (thermoset resin) as high performance compounds for pipe, sanitary ware, tank, and parts used in electrical, construction, transportation, and building industry have extensive application areas. High-thermal stability, flame retardant, good mechanical features, and low-shrinkage properties were expected from the technical products. Ecological-filling agents for achieving flame-retardant polymer composites were recently researched. It was requested that filling agent amount used to gain flame resistance to polymer composites should not deteriorate their mechanical properties. In this chapter, researchers investigated fire resistance, smoke emission, and thermal stability properties of nanocomposites. They were produced from unsaturated polyester and glass-reinforced polyester laminated from unsaturated polyester with the addition of multiingredients such as nanoparticles (montmorillonite, expandable graphite, and silica) and phosphate, borate, and tin-zinc compounds as filling agent instead of halogenated compounds and macroscopic particles and also compared the results with that of nanocomposites produced with the addition of one filling agent. Up to now, researches about composites mainly composed of polymers such as polystyrene, polyolefins, polyamides, PVC, and elastomers such as rubbers—SBR, BR, and

HNBR—as matrix material were carried out. In this chapter, the properties of nanocomposites produced from unsaturated polyester as matrix material and that contained multi-ingredients as filling agent were evaluated.

In Chapter 4, the use of polyester resins for automotive and aerospace applications such as panels, automotive dashboards, light covers, space ships, pressure vessels, water tanks, automotive driving shafts, rocket motor, and helicopter blades was disclosed. Moreover, their use in the production of these mechanical parts and assemblies and electrical insulation as well as the production methods of these components from polyester resins were presented. While producing the mechanical parts, production methods such as pultrusion, filament winding, wet lay-up method, autoclave curing and vacuum bagging, vacuum-assisted resin transfer, and resin transfer molding were used. The different products of electrical insulation from polyester resins were produced with various impregnation methods such as flood impregnation, vacuum pressure impregnation (VPI), vacuum impregnation (VI), trickle impregnation or dip and bake impregnation, wet windings, B-stage tapes, potting, and full encapsulating. The production methods of mechanical and electrical parts from polyester resins were examined under two groups, and their pros and cons were emphasized in this chapter. Lastly, production parameters affecting the quality and production were summarized.

In Chapter 5, the authors investigate the functional, mechanical, and thermal properties of glass fiber–reinforced polyester composites filled with various contents of TiO_2 particles. The composite materials were produced by hand lay-up process (HLUP) and compression molding technique (CMT). Moreover, the cross-linking mechanism of unsaturated polyester resin (UPR) with TiO_2 particles and glass fibers was extensively explained in the chapter. The fiber-reinforced composite materials (FRPCs) could be applied to automotive, marine, aerospace, construction, and sports industries. Imbedding of FRPCs with particles, such as metal oxides, improved the mechanical, functional, and thermal properties of FRPCs, and the composite materials are called polyester hybrid composites (PHCs).

Chapter 6 explains the textile products used in the automotive industry; the performance properties required from these textile materials; the places of usage in automotive industry such as carpets, seat covers, interior components, tires, filter, and safety equipment; and the importance and the use of polyester fiber in the application area when compared with that of the other fibers. Furthermore, the concern of recycling and reuse of polyester was raised. The researches about polyester and recycled polyester used in the automotive industry were reviewed. Commercial automotive textiles constructed from polyester were also exemplified and information about innovative applications and trends in the automotive textiles was given.

Nurhan Onar Çamlibel, PhD
Textile Engineering Department
Pamukkale University, Engineering Faculty
Denizli, Turkey

Chapter 1

Introductory Chapter: Introduction to "Polyester – Production, Characterization and Innovative Applications"

Nurhan Onar Camlibel

Additional information is available at the end of the chapter

http://dx.doi.org/10.5772/intechopen.74422

1. Introduction

Polyesters are one of the most important and most used polycondensation polymers and are derived from dicarboxylic acids (sometimes other acid types) and diols. Polyester is a polymer class containing ester functional group on polymeric main chain. Polyester term is usually used for polyethylene terephthalate (PET), despite the numerous polyester forms are present [1, 2].

Recent researches have indicated that using of polyester polymer processed into various forms, e.g., fibers, filament, fabric, composites, resins, dendrimers, films, sheets, and membranes in different fields, such as textile, automotive, medical, electronic, and construction applications, attracts worldwide interest. Polyester is also extensively used as packaging materials, such as bottle/containers. The polymerization of polyester could be carried out as polycondensation, ring-opening polymerization, and polyaddition processes. Furthermore, polyester could be recycled by physical (mechanical) or chemical (hydrolysis, methanolysis, and glycolysis reactions) methods [1, 2].

Recycled polyester could be used for packaging, construction parts, pipes, tanks, geotextiles, nonwoven, carpets, etc. It is expected to run out of crude oil reserves at World in 2043. Thus, recycling of petroleum-based polymers is crucial. In addition, effectively recycling of polyester will give rise to lessen carbon dioxide emission and thus global warming [1, 2].

Polyester can be classified into two groups: thermoplastic polyesters and thermoset (unsaturated polyester, polyester resin) polyesters. Thermoplastic polymers could be categorized as linear aromatic polyesters (fiber- and film-forming polyesters), elastomers (block copolyesters), liquid crystal polyester, engineering plastics, aliphatic polyesters, and poly(hydroxyl alkanoates) [1, 2].

2. Thermoset polyesters (unsaturated polyester)

Unsaturated polyester resins (UPRs) are used in civil/structural engineering applications, ships materials, composites, construction, piping, storage tanks, protective coatings, and automotive paints, which required high strength, ductility properties, and fire resistance [3–5]. The composite materials composed of unsaturated polyester are produced by some methods, such as lay-up method, pultrusion, filament winding, vacuum bagging and autoclave curing, and liquid molding [6]. Unsaturated polyester resins have advantages due to their chemical resistance, electrical properties, rapid curing, and relatively low prices for using the application areas. However, UPRs are extremely flammable and produce toxic smoke during combustion. Some filler and additives such as alumina trihydrate (ATH), magnesium hydroxide, nickel hydroxide, molybdenum disulphide, nanoclay, antimony oxide, gypsum particles, graphene, and carbon nanotube in the resins and some modifications of the resins with halogens such as bromine or phenolic resin could be used to improve fire resistance of the composites. Furthermore, their stiffness and strength properties were limited in comparison with other thermosetting resins. Fiber-reinforced unsaturated polyesters (UPs) and filler containing UPs were widely investigated to surpass these deficiencies [3–5]. Applications of unsaturated polyester resins were illustrated in **Figure 1**.

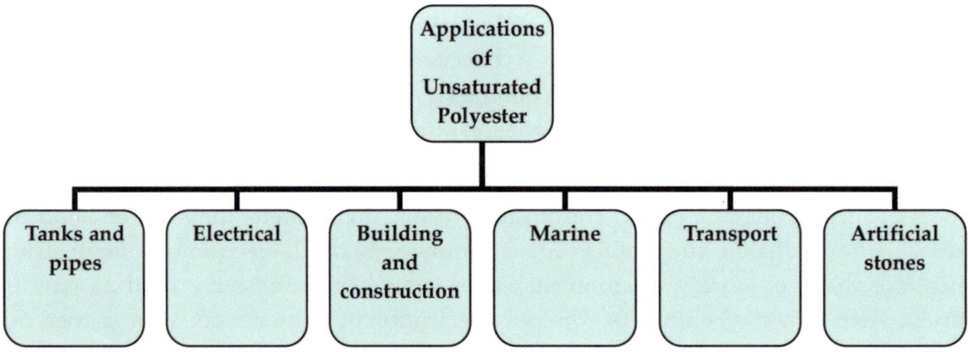

Figure 1. Applications of unsaturated polyester resins [7].

3. Thermoplastic polyesters

Polyester is the most produced thermoplastic polymer and has many application fields such as especially textile fibers and bottles. Polyester used in textiles is generally polyethylene terephthalate (PET), whose well-known properties are chemical inertness, lightness, good processability, high-melting point, high tenacity, and low cost [8, 9]. Furthermore, recycling of polyester bottles for sustainable textiles recently draws attention in the world and provides diverting waste as bottles from landfills, reducing environmental pollution and reducing carbon footprint and save energy compared to produce virgin polyester [10]. The bottle could be recycled by mechanical and chemical processes. Chemical process is based on the depolymerization of polyester to oligomers or monomers by chemical reaction, while mechanical process is carried out with melting and re-extruding to make fiber [11, 12].

Author details

Nurhan Onar Camlibel

Address all correspondence to: nonar@pau.edu.tr

Department of Textile Engineering, Faculty of Engineering, Pamukkale University, Denizli, Turkey

References

[1] Deopura BL, Alagirusamy R, Joshi M, Gupta B, editors. Polyesters and Polyamides. Vol. 71. England: Woodhead Publishing in Textiles, Cambridge; 2008. pp. 1-11

[2] Yamashita H, Nakano Y. Polyester: Properties, Preparaton and Applications. New York: Nova Science Publishers, Incorporated; 2008. pp. VII-X

[3] Tabatabai H, Janbaz M, Nabizadeh A. Mechanical and thermo-gravimetric properties of unsaturated polyester resin blended with FGD gypsum. Construction and Building Materials. 2018;**163**:438-445. DOI: 10.1016/j.conbuildmat.2017.12.041

[4] Hai Y, Jiang S, Qian X, Zhang S, Sun P, Xie B, Hong N. Ultrathin beta-nickel hydroxide nanosheets grown along multi-walled carbon nanotubes: A novel nanohybrid for enhancing flame retardancy and smoke toxicity suppression of unsaturated polyester resin. Journal of Colloid and Interface Science. 2018;**509**:285-297. DOI: 10.1016/j.jcis.2017.09.008

[5] Hashemi MJ, Jamshidi M, Aghdam JH. Investigating fracture mechanics and flexural properties of unsaturated polyester polymer concrete (UP-PC). Construction and Building Materials. 2018;**163**:767-775. DOI: 10.1016/j.conbuildmat.2017.12.115

[6] Campbell FC. Chapter 5: Thermoset composite fabrication processes. In: Structural Composite Materials. Ohio: ASM international; 2010. p. 119

[7] World Unsaturated Polyester Resin Market – Opportunities and Forecasts, 2017-2023. Available from: https://www.alliedmarketresearch.com/unsaturated-polyester-resin-market [Accessed: Jan 20, 2017]

[8] Wang X, Lu W, Chen Y, Li N, Zhu Z, Wang G, Chen W. Effective elimination of antibiotics over hot-melt adhesive sheath-core polyester fiber supported graphitic carbon nitride under solar irradiation. Chemical Engineering Journal. 2018;**335**:82-93. DOI: 10.1016/j.cej.2017.10.061

[9] Dave H, Ledwani L, Chandwani N, Kikani P, Desai B, Chowdhuri MB, Nema SK. Use of dielectric barrier discharge in air for surface modification of polyester substrate to confer durable wettability and enhance dye uptake with natural dye eco-alizarin. Composite Interfaces. 2012;**19**(3-4):219-229. DOI: 10.1080/15685543.2012.702594

[10] Patnaik A, Mvubu M, Muniyasamy S, Botha A, Anandjiwala RD. Thermal and sound insulation materials from waste wool and recycled polyester fibers and their biodegradation studies. Energy and Buildings. 2015;**92**:161-169. DOI: 10.1016/j.enbuild.2015.01.056

[11] Vadicherla T, Saravanan D. Thermal comfort properties of single jersey fabrics made from recycled polyester and cotton blended yarns. Indian Journal of Fibre and Textile Research. 2017;**42**(3):318-324

[12] Why is Recycled Polyester Considered a Sustainable Textile? Available from: http://oecotextiles.wordpress.com/2009/07/14/why-is-recycled-polyester-considered-a-sustainable-textile [Accessed: Jan 20, 2017]

Chapter 2

PET Bottle Recycling for Sustainable Textiles

Esin Sarioğlu and Hatice Kübra Kaynak

Additional information is available at the end of the chapter

http://dx.doi.org/10.5772/intechopen.72589

Abstract

Polyethylene terephthalate (PET), as the most favorable packaging material, is owing to its transparent color, lightweight, strength, food safe, inexpensive price, fully recyclability, etc. In addition to all these advantages, PET as a waste material takes up considerable space in nature and needs to be recycled for the disposal of these wastes. In this regard, recycling enables conserving raw materials, reducing energy use in order to produce virgin PET, and reducing greenhouse gas emissions. Today, PET is the most widely recycled plastic in the world. Eco-friendly products obtained by recycling of PET are mainly used as textile fibers. In addition, both brands and consumers are seen to be enthusiastic in order to minimize the environmental effects of PET wastes. This study is concerned with the use of textile fiber from recycled PET (r-PET) bottles to produce a cotton blended ring and compact yarns. Undoubtedly, the study also includes comparison of cotton blended virgin polyester fiber (v-PET) with r-PET fiber to determine the advantages and disadvantages of r-PET fiber. The reason for choosing cotton fiber is the most preferred fiber blending with PET commercially.

Keywords: r-PET, recycle, PET bottle, ring spinning system, compact spinning system, virgin PET

1. Introduction

Economic and population growth and industrialization in the world together cause an increase in the amount of waste. As a consequence of all these, while the more intensive use of natural resources is inevitable, the wastes created by the ever-increasing consumption tendency have reached the huge amounts that threaten the environment and human health due to their quantity and harmful contents. For this purpose, waste policies should be developed and waste management studies should also be carried out, especially in the field of recycling these wastes, because of long decomposition time of these wastes in the environment causing landfill problem [1, 2].

Waste management system enables collection, categorization, reduction, recycling, and reuse of waste. At present, countries' intensive efforts on waste management are striking. Waste management, which has an important place among environmental protection policies, should prevent the rapid depletion of natural resources and minimize the potential risks of the wastes to the environment and human health [3].

Polyethylene terephthalate (PET) is a versatile material and has a wide range of applications such as clothing, acoustic panels, sportswear, agricultural nets, nonwovens, sheets and films, straps, engineering resins, food and beverage bottles, bottles, packaging materials, reinforcement in building construction, etc. Among these products, bottle grade PET is generally used for water and beverage packaging due to its lightweight, inexpensive price, resistance to microorganisms, and light [4–6]. Bottles of water, soft drinks, and other beverages constitute 83–84% of global PET resin requirement [7]. Furthermore, the projected demand for PET packaging materials is forecasted to reach 20 million tons by 2019 with an annual increase of 4.6% [8].

With the widespread application of PET, large quantities of PET waste were inevitably created. PET has no side effects on the human body and does not pose a direct threat to the environment. On the other hand, it is regarded as a harmful material because of its high volumetric fraction in the waste stream and high resistance to atmospheric and biological agents [9].

Due to poor biodegradation of PET, it is difficult to remove waste. It is possible to suggest two acceptable solutions; burning and recycling. Burning method arises releasing toxic fumes into the atmosphere, causing environmental pollution and health risks [4, 10]. As an acceptable solution, the recycling of PET bottles enables the conservation of natural sources such as fossil fuels and energy, solving landfill problem, reducing greenhouse gas emission, lowering carbon footprint, creating new business opportunities as well as a contribution to the national economy [7, 11–14]. In addition, recycling processes are the best way to economically reduce PET waste [15]. With both reduced energy costs and raw material costs, recycling fiber production has become a form of production with a significant economic advantage [16]. Two forms of PET bottle recycling can be distinguished as a closed loop and open loop recycling. Closed loop recycling or bottle-to-bottle refers to a product system that recycles post-consumer waste within the same system. Open loop recycling denotes the utilization of recycled material in another product system such as bottle-to-fiber recycling [16–18]. **Figure 1** displays the process of bottle-to-fiber recycling.

PET flakes are obtained from PET bottle wastes after a series of procedures such as sorting, washing, grinding, drying, etc. [19]. Most of the recycled PET flakes produced worldwide are utilized for staple fiber applications in textile sector (**Figure 2**) [8, 20, 21]. Because of environmental reasons initially, the recycling of PET bottles to textile fibers has now become commercially attractive [22]. Furthermore, as petroleum prices increase, recycling of PET becomes more financially feasible rather than a virgin PET. It is expected that the recycling of the PET bottle will be estimated up to annually 13 million tons in 2018 and up to 15 million tons in 2020 [21].

The fiber obtained from PET flakes has dominant proportion among end users as 44% of total market share in 2016. These fiber materials are generally called as recycled PET (r-PET) and especially used in carpets, blankets, clothing, and other textile applications [12, 14, 23]. However,

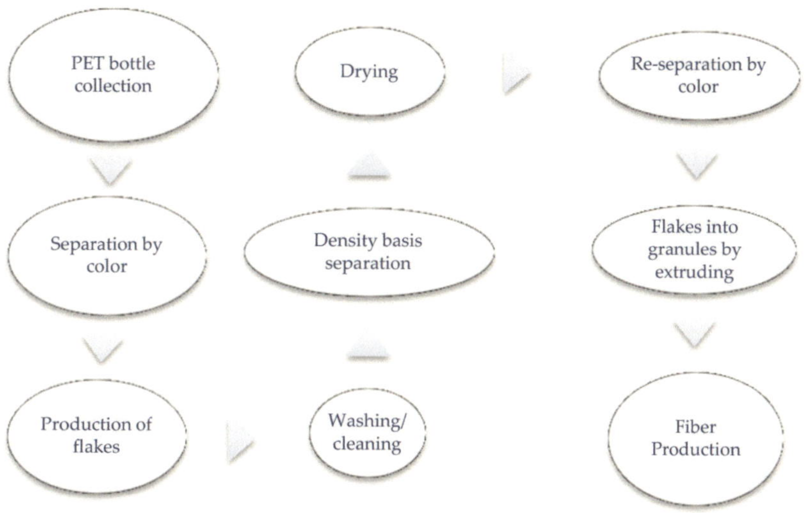

Figure 1. Bottle-to-fiber recycling.

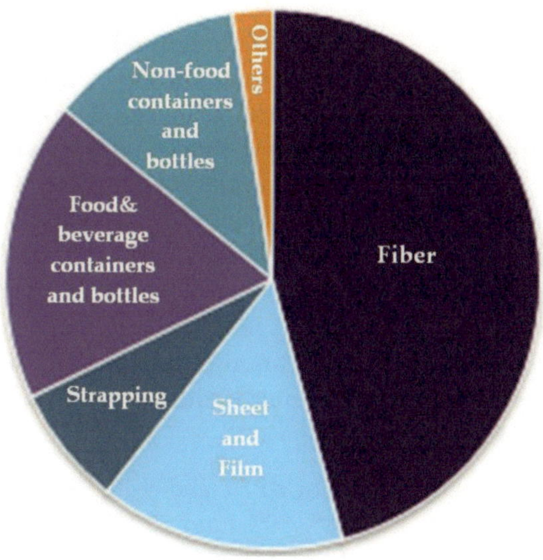

Figure 2. Global annual r-PET market volume by end-use in 2016 [23].

properly sorting PET bottle wastes and carefully removing impurities are essential so as to obtain similar recycled fiber quality as virgin ones [24]. Furthermore, there are significant limitations in the use of r-PET for the production of partially oriented yarn (POY), drawn textured yarn (DTY) and fully drawn yarn (FDY) type textile yarns, microfilaments, tire cord or high quality biaxial films [21].

Converting of PET flakes into PET fibers can be carried out by two main processes, such as chemical recycling and mechanical recycling. The chemical recycling method provides value-added products from PET bottle wastes, and depolymerization of PET by hydrolysis, methanolysis, and glycolysis is used to re-use regenerated raw materials as monomers for new polymerization processes [10, 12, 16, 25–27]. The chemical recycling technique produces superior quality materials, but this method is highly labor and power-intensive, so it requires high processing costs [4, 12, 15, 21].The mechanical recycling method involves sorting and separation of waste, washing for removal of dirt and contaminants, grinding in order to obtain flakes, cleaning, separating, dehydrating, drying, and re-melting [10, 16, 21, 25, 27]. Mechanical recycling is preferred due to a significant reduction in processing costs, global warming potential, non-renewable energy use, abiotic exhaustion, acidification, eutrophication, human toxicity, and water toxicity [4].

As such, studies on the use of r-PET staple fiber in the field of textile applications have established the focus of researchers. Yüksekkaya *et al.* investigated the properties of yarns and knitted fabrics produced by virgin PET and r-PET and cotton fibers as virgin and recycled. Yarns were produced as 100% virgin, 100% recycled, and 50%/50% virgin/recycled proportion by using rotor spinning machine. They stated that yarns produced from recycled fibers had better yarn unevenness, lower number of yarn imperfections, and better yarn quality index value. In addition, yarn tensile strength and knitted fabric burst strength were found to be lower for recycled yarns and fabrics when compared to virgin ones [28].

Another study was conducted on the performance and durability of woven fabrics made from r-PET with different ratios by Mari and Shinji. Commercially available plain and twill woven samples were collected and categorized according to the recycling content. They observed that more fatigue action occurred for fabrics including r-PET than virgin PET after washing. It was found that fabrics with r-PET exhibited higher stiffness with increasing r-PET content [29]. Rajamanickam and Vasudevan studied on the antibacterial activities of 19.7 tex ring spun yarns including lyocell and r-PET at different blending ratios (100% r-PET, 70%/30%, 50%/50% lyocell/r-PET, and 100% lyocell). Before the antibacterial activities, yarns were treated with chitosan finish. In this study, it was suggested that the yarns can be used as hospital textiles and blended yarn samples exhibited better antimicrobial activity. Furthermore, "blends of lyocell & r-PET yarns" were found to be suitable in the field of hospital textiles due to having higher tenacity and elongation property [30].

Telli and Özdil studied the performance of r-PET and v-PET blended yarns with cotton at different blend ratios (100%, 70%/30%, 50%/50%, and 30%/70%). In that respect, 30 tex yarn samples were manufactured by means of ring spinning system and tensile properties, unevenness, imperfection index (IPI), and hairiness of these yarns were determined. In this study, it is suggested to encourage the consumption of the lower cost r-PET fibers having current quality standards in this field [31]. Another study of the authors is related to the knitted fabric performance made from these yarns. It was found that knitted fabrics including virgin PET had better performance specifications than that of r-PET. According to the test results, they offered to use blends of r-PET fibers in suitable proportions according to usage area in order to prevent the decrease in fabric performance instead of using virgin PET fiber [19].

In order to contribute the use of r-PET staple fiber in the textile sector, producers should be encouraged and some legislation should be implemented. Since recycling of PET bottles is an important value-added sector; it prevents the consumption of limited fuel source and energy so it helps to protect the environmental pollution and decreases the landfilling problem. There are various attempts of big textile brands about using of r-PET fibers in their products such as Zara, H&M, Nike, Adidas, and Lewis® etc. to fulfill their social responsibilities to nature. With this study, we aimed to determine the performance of r-PET containing yarns produced with different spinning systems at the same production conditions. For this purpose, r-PET fibers obtained from PET flakes by mechanical recycling methods were used and blended with cotton fiber at different ratios. The yarn samples with the same blend ratios were produced with v-PET fiber instead of r-PET, in order to compare the performance of recycled and virgin fibers. Two spinning systems were selected as ring and compact which are commercially used in industry. Tensile properties, unevenness, IPI, hairiness of these yarns were determined and compared. IBM® SPSS® 20 statistical package program was used to determine the statistical significance of the effects of raw material, spinning system and blend ratio on yarn performance characteristics.

2. Materials and methods

2.1. Materials

In this research, authors cooperated with Gama Recycle Textile Company in Gaziantep/Turkey. PET flakes were manufactured by collecting, rating, washing and grinding, drying and the size of PET bottle wastes. Then, PET flakes were converted into r-PET fiber by a mechanical process. The most commercially preferred r-PET fiber was chosen among the productions of the company. r-PET ,v-PET, and cotton (Co) fibers were used as raw materials for yarn production (**Table 1**).

Fibers were weighted and blended with sandwich blend method at the beginning of the blowroom. In the yarn production stage, blowroom, carding, three passage drawframe, and roving processes were achieved. In addition, 19.7 tex ring and compact spun yarns were produced with α_e = 3.39 (730 TPM) twist level at different blend ratios (70%/30%, 50%/50%, and 30%/70%) of r-PET and v-PET with cotton fiber, separately. The spindle speed was constant at 13,500 rpm. The 100% r-PET and 100% v-PET ring and compact yarns were also produced with the same production parameters to compare and evaluate the differences in yarn properties systematically. Nevertheless, the production of the 100% Co ring and compact yarns were conducted with α_e = 3.71 (800 TPM) twist level at the same spindle speed, owing to the necessity of producing all yarn samples at optimum process parameters [32]. Since the

	r-PET	v-PET	Co
Fiber linear density (dtex)	1.30	1.30	1.70*
Length (mm)	38	32	32

*4.32 Micronaire (dtex Cotton = Micronaire*0.394).

Table 1. Raw material properties.

higher twist was necessary to hold the fibers together and to obtain required cohesion forces between fibers in the production of the 100% Co ring and compact yarns.

2.2. Methods

Yarn samples were conditioned in standard atmosphere conditions at 20 ± 2°C temperature and 65 ± 4% relative humidity for 24 hours [33]. Tensile properties were carried out by Uster® Tensorapid-4 according to the related standard with 10 measurements from every four bobbins [34]. Five tests from each four bobbins were carried out in order to determine unevenness, IPI, and hairiness of yarn samples with Uster® Tester 4 in accordance with ISO standard [35].

In statistical analysis, multivariate analysis of variance (MANOVA) was achieved at 95% confidence interval by means of IBM® SPSS® 22.0 package program to compare whether there is statistically significant effect of the independent parameters as raw material (r-PET and v-PET), blend ratio (100%, 70%/30%, 50%/50%, 30%/70% and 0%), and yarn type (ring and compact) on yarn unevenness, IPI, hairiness and tensile properties. In addition, multiple comparison tests were obtained to determine the difference between two groups via pair wise comparison at a 0.05 significance level.

3. Experimental results

3.1. Unevenness

CVm% is defined as unevenness of yarn samples and is shown in **Figure 3**. When the raw material is considered, lower CVm % values are observed for pure v-PET yarn samples for both yarn types.

On the other hand, the maximum CVm % was recorded for 100% Co ring spun yarn. There is a tendency of increasing in unevenness from 100% pure synthetic yarn to 100% Co yarn. It is

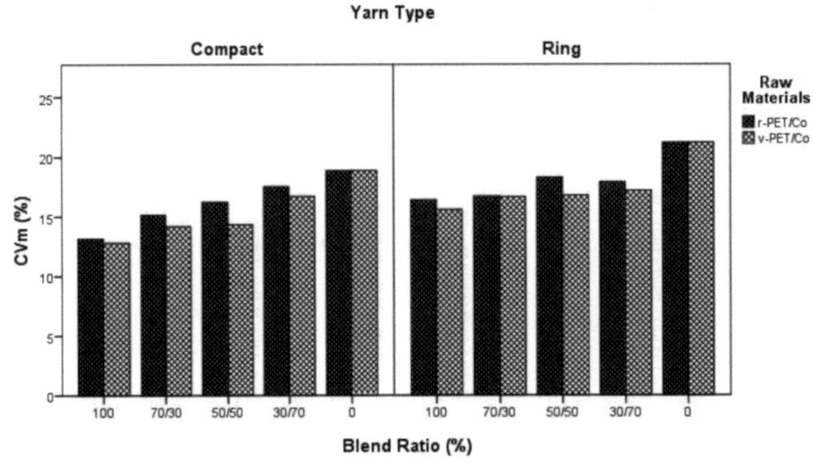

Figure 3. CVm % of yarn samples.

probable that the incorporation of synthetic fibers having a more uniform fiber structure than the Co fiber in the yarn cross-section reduced the unevenness. Comparison in between the ring and compact yarn types, ring spun yarns for all samples have higher unevenness owing to the higher number of protruding fiber ends from yarn body. When results are examined generally, r-PET raw material cause higher yarn unevenness than v-PET. It should also be noted that higher yarn unevenness will cause lower tensile strength and higher IPI. Consequently, it is preferable to use 70%/30% blend ratio and compact spinning system for a better yarn unevenness, aiming using the r-PET raw material.

3.2. IPI

Imperfections are referred to as frequently occurring yarn faults as percentage. Imperfection index in the yarn refers to the total number of thin places (−50%), thick places (+50%) and neps (+200%) of yarn in kilometer. IPI values of yarn samples are given in **Figure 4**.

According to IPI values, it is seen that v-PET blended yarns have lower values than that of r-PET ones. On the other hand, for all blended samples, it is seen that increasing PET content cause a decrease in IPI values. Furthermore, 100% r-PET yarn samples exhibit similar IPI values with 100% v-PET samples. Compact spinning technology has a decreasing effect on IPI for all samples. Consequently, it can be concluded that higher IPI values of r-PET blended yarns may be a result of low compatibility of r-PET fibers with Co fiber. Higher IPI values of r-PET blended yarns should be compensated by using higher PET ratio and preferring compact spinning technology.

3.3. Hairiness

From **Figure 5**, lower hairiness values of compact spun yarn samples than that of ring ones are obvious. Since in the compact spinning system, it is aimed to provide a less hairy yarn by contributing the protruding fiber ends into yarn body via reforming the spinning triangle. On the other hand, it is seen that r-PET samples have slightly higher hairiness values than that of v-PET samples for ring spun yarns, whereas there is almost no difference between r-PET and

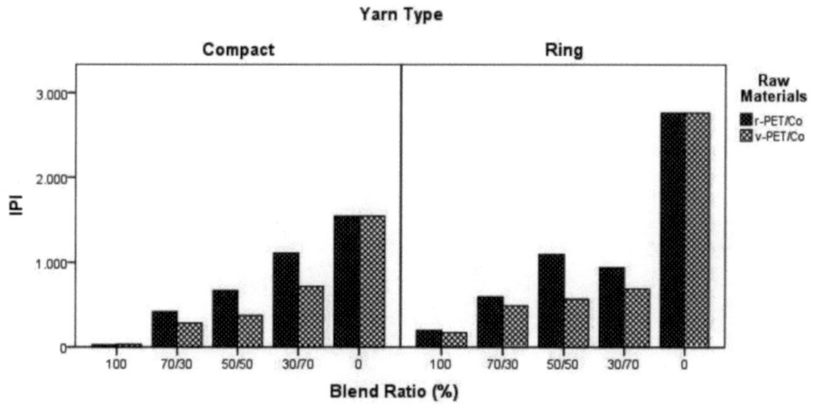

Figure 4. IPI of yarn samples.

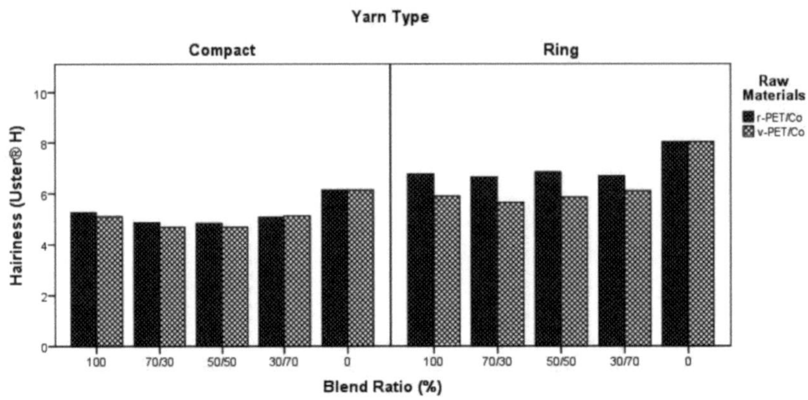

Figure 5. Hairiness of yarn samples.

v-PET for compact spun yarns. Among blended yarn samples of both r-PET and v-PET fiber types, it is clear that the lowest yarn hairiness is obtained with 50%/50% blend ratio.

3.4. Tensile properties

Figure 6 displays tensile strength values of r-PET/Co and v-PET/Co ring and compact yarns at different blend ratios. It is obviously seen that v-PET containing yarns have higher tensile strength than that of r-PET ones and 100% v-PET compact yarn has the highest strength. This is an expected situation owing to the fact that r-PET fiber has lower fiber strength due to recycling. On the other hand, for ring spun yarn type, 100% v-PET yarn has higher strength value than that of 100% r-PET yarn about 28% at same production parameters. 100% v-PET compact spun yarn has higher tensile strength with the value of 39% than 100% r-PET compact yarn. This is an obvious positive effect of the compact spinning system on yarn strength. But also it should be considered that the magnitude of this effect is lower for PET fibers than v-PET.

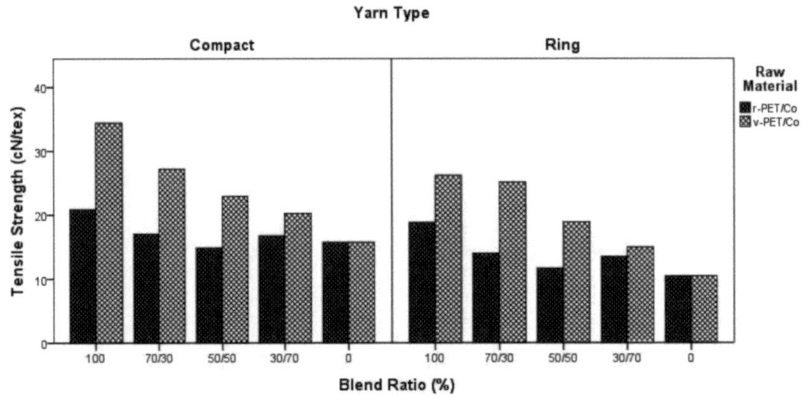

Figure 6. Tensile strength of yarn samples.

Having higher tensile strength is an important issue for both fabrication processes and fabric performance during use life. So that, 70%/30% blend ratio and compact spinning technology should be preferred for a higher yarn strength. When yarn type is taken into consideration, as an expected result is that the tensile strength of the compact spun yarns are higher than that of conventional ring yarns [36–41]. Since compact spinning technology contributes yarn evenness by increasing the number of protruding fiber ends into yarn body. This reforming in yarn body also contributes yarn tenacity by increasing the number of fibers in yarn cross-section that withstand against tensile force. On the other hand, it is also clearly seen that r-PET and v-PET presence contribute yarn tensile strength with respect to the Co content. The bar graphs for elongation of yarn samples are illustrated in **Figure 7**.

Pure r-PET ring and compact spun yarns exhibit a higher elongation at break than a pure v-PET ring and compact yarns, it may be probably due to the fact that higher elongation of r-PET fiber than that of v-PET. It is determined that both the compact and the ring spun yarns elongation at break decrease with increasing Co fiber ratio.

3.5. Statistical analysis

MANOVA analysis was carried out to determine the significance of the effects of raw material, yarn type and blend ratio on tensile properties, unevenness, imperfections and hairiness of yarn samples (**Table 2**). Results show that yarn type, raw material, and blend ratio has significant effects on all response variables at 0.05 level, except the insignificant effect of the raw material on yarn elongation. Furthermore, there is no statistically significant difference between raw materials in elongation of r-PET or v-PET containing ring and compact yarns. R^2 value is defined as the magnitude of the effect of the independent variables on the response variables in percent. The strength of the relationship between independent variables and response variables are explained in the range of 0-100%. The higher the R^2 means the better the model fits your data.

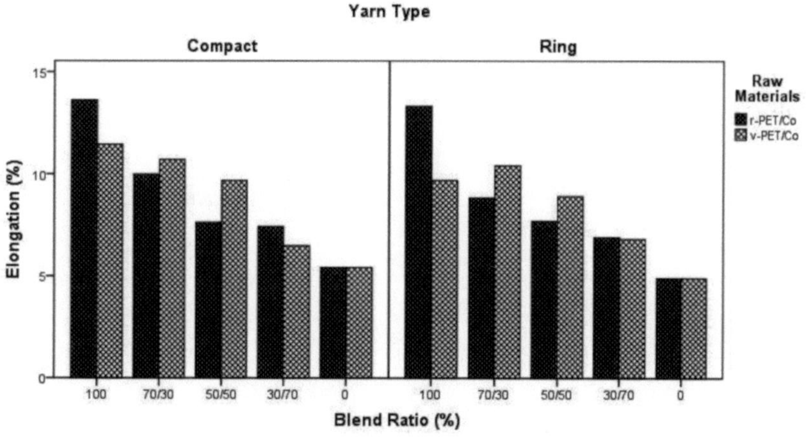

Figure 7. Elongation of yarn samples.

Source	Dependent variable	Sum of squares	df	Mean square	F	Sig.
Corrected model	Tensile strength	2904.574 **(a)**	19	152.872	75.864	**0.000***
	Elongation	499.457 **(b)**	19	26.287	57.934	**0.000***
	CVm	344.032 **(c)**	19	18.107	63.940	**0.000***
	IPI	35363708.784 **(d)**	19	1861247.831	50.887	**0.000***
	Hairiness	87.003 **(e)**	19	4.579	165.571	**0.000***
Intercept	Tensile strength	27369.271	1	27369.271	13582.285	**0.000***
	Elongation	5916.800	1	5916.800	13039.899	**0.000***
	CVm	22291.832	1	22291.832	78717.571	**0.000***
	IPI	50949906.153	1	50949906.153	1392.992	**0.000***
	Hairiness	2812.836	1	2812.836	101707.072	**0.000***
Yarn type (A)	Tensile strength	347.403	1	347.403	172.402	**0.000***
	Elongation	9.884	1	9.884	21.784	**0.000***
	CVm	72.048	1	72.048	254.418	**0.000***
	IPI	1781597.278	1	1781597.278	48.710	**0.000***
	Hairiness	46.772	1	46.772	1691.195	**0.000***
Raw material (B)	Tensile strength	777.442	1	777.442	385.814	**0.000***
	Elongation	0.030	1	0.030	.065	0.799
	CVm	9.099	1	9.099	32.131	**0.000***
	IPI	553363.278	1	553363.278	15.129	**0.000***
	Hairiness	2.915	1	2.915	105.389	**0.000***
Blend ratio (C)	Tensile strength	1347.936	4	336.984	167.232	**0.000***
	Elongation	434.549	4	108.637	239.423	**0.000***
	CVm	240.446	4	60.111	212.267	**0.000***
	IPI	29589733.206	4	7397433.302	202.249	**0.000***
	Hairiness	30.965	4	7.741	279.912	**0.000***
A * B	Tensile strength	13.082	1	13.082	6.492	**0.013***
	Elongation	0.964	1	0.964	2.124	0.150
	CVm	0.364	1	0.364	1.287	0.261
	IPI	1432.278	1	1432.278	.039	0.844
	Hairiness	1.821	1	1.821	65.846	**0.000***
A * C	Tensile strength	20.552	4	5.138	2.550	**0.048***
	Elongation	1.206	4	0.301	.664	0.619
	CVm	14.704	4	3.676	12.981	**0.000***
	IPI	2797362.394	4	699340.598	19.120	**0.000***
	Hairiness	3.086	4	0.772	27.900	**0.000***
B * C	Tensile strength	367.084	4	91.771	45.542	**0.000***

Source	Dependent variable	Sum of squares	df	Mean square	F	Sig.
	Elongation	49.650	4	12.413	27.356	**0.000***
	CVm	6.047	4	1.512	5.339	**0.001***
	IPI	557862.644	4	139465.661	3.813	**0.008***
	Hairiness	0.959	4	0.240	8.669	**0.000***
A * B * C	Tensile strength	31.075	4	7.769	3.855	**0.007***
	Elongation	3.175	4	0.794	1.749	0.151
	CVm	1.323	4	0.331	1.168	0.334
	IPI	82357.706	4	20589.427	.563	0.690
	Hairiness	0.484	4	0.121	4.374	**0.004***
Error	Tensile strength	120.904	60	2.015		
	Elongation	27.225	60	0.454		
	CVm	16.991	60	0.283		
	IPI	2194552.313	60	36575.872		
	Hairiness	1.659	60	0.028		
Total	Tensile strength	30394.749	80			
	Elongation	6443.482	80			
	CVm	22652.855	80			
	IPI	88508167.250	80			
	Hairiness	2901.498	80			
Corrected total	Tensile strength	3025.478	79			
	Elongation	526.682	79			
	CVm	361.023	79			
	IPI	37558261.097	79			
	Hairiness	88.662	79			

(a) $R^2 = 96$ (Adjusted $R^2 = 94.7$), (b) $R^2 = 94.8$ (Adjusted $R^2 = 93.2$), (c) $R^2 = 95.3$ (Adjusted $R^2 = 93.8$), (d) $R^2 = 94.2$ (Adjusted $R^2 = 92.3$), and (e) $R^2 = 98.1$ (Adjusted $R^2 = 97.5$). *The mean difference is significant at 0.05 level.

Table 2. Multivariate analysis of variance (MANOVA) statistical analysis results.

Multiple comparison results provide to analyze the differences between two samples differently from MANOVA providing to see only if there is a statistically significant difference among sample groups. The results in **Table 3** involve r-PET and v-PET samples to focus on the effect of blend ratio for all samples. According to multiple comparison tests, it is seen that the difference between sample groups are generally important (α at 0.05 significance level). For tensile strength, it is seen that 70%/30% and 50%/50% yarn samples provide statistically similar results. In other words, these two samples compose a group regarding the tensile strength. Also, the rest of the samples have statistically different strength results among each other. On the other hand, all of the samples have statistically different yarn elongation, CVm % and IPI

		Dependent variable				
		p Value				
Blend ratio	Blend ratio	Tensile strength (cN/tex)	Elongation (%)	CVm (%)	IPI (%)	Hairiness (Uster® H)
0	100	0.000*	0.000*	0.000*	0.000*	0.000*
	30/70	0.000*	0.000*	0.000*	0.000*	0.000*
	50/50	0.000*	0.000*	0.000*	0.000*	0.000*
	70/30	0.000*	0.000*	0.000*	0.000*	0.000*
100	0	0.000*	0.000*	0.000*	0.000*	0.000*
	30/70	0.000*	0.000*	0.000*	0.000*	0.930
	50/50	0.000*	0.000*	0.000*	0.000*	0.004*
	70/30	0.000*	0.000*	0.000*	0.000*	0.000*
30/70	0	0.000*	0.000*	0.000*	0.000*	0.000*
	100	0.000*	0.000*	0.000*	0.000*	0.930
	50/50	0.593	0.000*	0.000*	0.041*	0.041*
	70/30	0.000*	0.000*	0.000*	0.000*	0.001*
50/50	0	0.000*	0.000*	0.000*	0.000*	0.000*
	100	0.000*	0.000*	0.000*	0.000*	0.004*
	30/70	0.593	0.000*	0.000*	0.041*	0.041*
	70/30	0.000*	0.000*	0.008*	0.013*	0.668
70/30	0	0.000*	0.000*	0.000*	0.000*	0.000*
	100	0.000*	0.000*	0.000*	0.000*	0.000*
	30/70	0.000*	0.000*	0.000*	0.000*	0.001*
	50/50	0.000*	0.000*	0.008*	0.013*	0.668

*The mean difference is significant at 0.05 level.

Table 3. Multiple comparison test results of yarn properties via blend ratio.

values among each other. With respect to yarn hairiness, 30%/70% samples and 100% samples have statistically similar values and on the other hand, 50%/50% samples have similar values with 70%/30% samples.

4. Conclusion

As a conclusion, it can be stated that r-PET fiber has enough contribution to be used with cotton fiber as a blended yarn component. IPI, unevenness, hairiness, and elongation performances of the r-PET yarn are at acceptable levels when they are compared with the v-PET yarns. The lower strength of the r-PET yarn may seem to be a disadvantage. But it must be encouraged as a social responsibility to use r-PET fiber in textile products in a certain proportion to maintain the nature, to recycle and to reuse PET bottle wastes in different application areas. Thus, it enables for reducing the volume of the landfill pollution, natural resources consumption, and extra energy costs required for virgin product handling.

Studies on the use of r-PET fibers are generally on the comparison of this fiber with r-PET fibers and using r-PET fibers in the blends with v-PET and Co fibers. The researchers emphasized the importance of using r-PET fiber in the textile industry as a result of these studies. However, some quality parameters require improvement in fiber properties. In systems that enable to recycle the PET bottle waste entered our lives as a part of waste management, mechanical and chemical works should be done for a better r-PET production with very close quality values with v-PET. Separate collection of PET bottles, proper categorization, and separation are required to capture sustainable r-PET fiber quality. In Turkey, while production of short fibers from PET flakes are intensive, with technological developments are now showing that providing filament from PET flakes production is also starting as investments.

In general, we can summarize the results as follows:

- v-PET containing yarns have higher tensile strength than that of r-PET ones and 100% v-PET compact yarn has the highest strength. This situation is expected since r-PET fiber has lower strength due to recycling and reforming processes.
- IPI and unevenness quality parameters are also very important in order to reason the usability of the yarn as a textile product. When the unevenness test results are evaluated, it is seen that the lowest CVm % values are obtained with pure v-PET yarn for both yarn spinning technologies. It should also be considered that higher yarn unevenness values caused higher IPI and lower tenacity.
- As a result of MANOVA analysis, yarn spinning technology, raw material and blend ratio generally have significant effects on response variables.

Acknowledgements

The authors would like to thank Gama Recycle Company/Gaziantep/Turkey for supplying r-PET fiber and contribution in the production of ring and compact yarn samples of this study. The authors also thank Karacasu Textile Company/Kahramanmaraş/Turkey for supplying v-PET fiber.

Author details

Esin Sarioğlu* and Hatice Kübra Kaynak

*Address all correspondence to: sarioglu@gantep.edu.tr

Department of Textile Engineering, Faculty of Engineering, Gaziantep University, Gaziantep, Turkey

References

[1] Tayyar AE, Üstün S. Usage of recycled PET. Pamukkale University. Journal of Engineering Sciences. 2010;**16**:53-62

[2] European Commission DG ENV. Plastic waste in the environment [Internet]. 2011. Available from: http://ec.europa.eu/environment/waste/studies/pdf/plastics.pdf [Accessed: October 8, 2017]

[3] Turkish Republic Ministry of Science, Industry and Technology General Directorate of Industry. National recycling strategy document and action plan 2014–2017. [Internet]. 2015. Available from: http://www.sanayi.gov.tr/DokumanGetHandler.ashx? Dokuman Id=19b39a64-ed35-4485-89df-aefea5bc21ea [Accessed: September 18, 2017]

[4] Barman NK, Bhattacharya SS, Mandot A. Mechanical properties of melt-spun monofilaments produced from virgin and recycled poly(ethylene terephthalate) blends. International Journal of Recent Scientific Research. 2015;**6**:4517-4525

[5] Luijsterburg BJ. Mechanical recycling of plastic packaging waste. Eindhoven: Technische Universiteit Eindhoven [thesis]; 2015. DOI: 10.6100/IR783771

[6] Shafique U, Zaman W, Anwar J, Munawar MA, Salman M, Dar A, Rehman R, Ashraf U, Ahmad S. A rapid, economical, and eco-friendly method to recycle terephthalic acid from waste poly (ethylene terphthalate) bottles. International Journal of Polymeric Materials. 2011;**60**:1147-1151. DOI: 10.1080/00914037.2011.557802

[7] Zhang H, Wen ZG. The consumption and recycling collection system of PET bottles: A case study of Beijing, China. Waste Management. 2014;**34**:987-998

[8] Sangalang A, Seok S, Kim DH. Practical design of green catalysts for PET recycling and energy conversion. In: Norena LE, Wang JA, editors. Advanced Catalytic Materials—Photocatalysis and Other Current Trends. Croatia: Intech; 2016. p. 141-166. DOI:10.5772/62041.ch.6

[9] Bartolome L, Imran M, Cho BG, Al-Masry WA, Kim DH. Recent developments in the chemical recycling of PET. In: Achilias DS, editors. Material Recycling—Trends and Perspectives. Croatia: Intech; 2012. p. 65-84. DOI: 10.5772/33800

[10] Park SH, Kim SH. Poly (ethylene terephthalate) recycling for high value added textiles. Fashion and Textiles. 2014;**1**:1-17

[11] Shen L, Nieuwlaar E, Worrell E, Patel MK. Life cycle energy and GHG emissions of PET recycling: Change-oriented effects. International Journal of Life Cycle Assessment. 2011;**16**:522-536. DOI: 10.1007/s11367-011-0296-4

[12] Dutt K, Soni RKA. Review on synthesis of value added products from polyethylene terephthalate (PET) waste. Polymer Science. 2013;**55**:430-452

[13] Hopewell J, Dvorak R, Kosior E. Plastics recycling: Challenges and opportunities. Philosophical Transactions of the Royal Society B. 2009;**364**:2115-2126. DOI: 10.1098/rstb.2008.0311

[14] Greene JP. Sustainable Plastics with Reduced Carbon Footprint & Reduced Waste [Internet]. Available from:http://www.sustainablegreenproducts.org/files/sustn-plastics.pdf [Accessed: October 1, 2017]

[15] Awaja F, Pavel D. Recycling of PET. European Polymer Journal. 2005;**41**:1453-1477

[16] Telli A, Özdil N, Babaarslan O. Usage of PET bottle wastes in textile industry and contribution to sustainability. Journal of Textiles and Engineer. 2012;**19**:49-55. DOI: 10.7216/130075992012198607

[17] Kuczenski B, Geyer R. Life Cycle Assessment of Polyethylene Terephthalate (PET) Beverage Bottles Consumed in the State of California [Internet]. 2014. Available from: http://www.calrecycle.ca.gov/publications/Documents/1487/20141487.pdf [Accessed: August 16, 2017]

[18] Plastics – the Facts 2014/2015 [Internet]. 2015. Available from: http://www.plasticseurope.org/documents/document/20150227150049final_plastics_the facts_2014_2015_260215.pdf [Accessed: October 10, 2017]

[19] Telli A, Özdil N. Effect of recycled PET fibers on the performance properties of knitted fabrics. Journal of Engineered Fibers and Fabrics. 2015;**10**:47-60

[20] Leonas KK. The use of recycled fibers in fashion and home products. In: Senthilkannan MS, editor. Textiles and Clothing Sustainability Recycled and Upcycled Textiles and Fashion. 1st ed. Singapore: Springer; 2017. pp. 55-77. DOI: 10.1007/978-981-10-2146-6_2

[21] Aizenshtein EM. Bottle wastes – to textile yarns. Fibre Chemistry. 2016;**47**:343-347. DOI: 10.1007/s10692-016-9691-8

[22] Upasani PS, Jain AK, Save N, Agarwal US, Kelkar AK. Chemical recycling of PET flakes into yarn. Journal of Applied Polymer Science. 2011;**123**:520-525. DOI: 10.1002/app.34503

[23] Recycled polyethylene terephthalate (PET) market analysis by product type (clear, colored), by end-use (fiber, sheet & film, strapping, f&b containers, bottles, non-food containers), and segment forecasts, 2014–2025 [Internet]. 2017. Available from: http://www.grandviewresearch.com/industry-analysis/recycled-polyethylene-terephthalate-pet-market [Accessed: October 16, 2017]

[24] Shen L, Worrell E, Patel MK. Open-loop recycling: A LCA case study of PET bottle-to-fibre recycling. Resources, Conservation and Recycling. 2010:1-19. DOI: 10.1016/j.resconrec.2010.06.014

[25] Lee JH, Lim KS, Hahm WG, Kim SH. Properties of recycled and virgin poly(ethylene terephthalate) blend fibers. Journal of Applied Polymer Science. 2013:1250-1256. DOI: 10.1002/APP.38502

[26] Elamri A, Lallam A, Harzallah O, Bencheikh L. Mechanical characterization of melt spun fibers from recycled and virgin PET blends. Journal of Materials Science. 2007;**42**:8271-8278. DOI: 10.1007/s10853-007-1590-1

[27] Koo HJ, Chang GS, Kim SH, Hahm WG, Park SY. Effects of recycling processes on physical, mechanical and degradation properties of PET yarns. Fibers and Polymers. 2013;**14**:2083-2087. DOI: 10.1007/s12221-013-2083-2

[28] Yüksekkaya ME, Celep G, Doğan G, Tercan M, Urhan BA. Comparative study of physical properties of yarns and fabrics produced from virgin and recycled fibers. Journal of Engineered Fibers and Fabrics. 2016;**11**:68-76

[29] Mari I, Shinji Y. Performance and durability of woven fabrics including recycled polyester fibers. Journal of Textile Engineering. 2004;**50**:25-30. DOI: 10.4188/jte.50.25

[30] Rajamanickam S, Vasudevan K. Study of antibacterial activity of chitosan on lyocell and recycled polyester yarns. International Journal of Innovative Research in Science Engineering and Technology. 2014;**3**:9480-9486

[31] Telli A, Özdil N. Properties of the yarns produced from r-pet fibers and their blends. Tekstil ve Konfeksiyon. 2013;**23**:3-10

[32] Sarıoğlu E. Ecological approaches in textile sector: The effect of r-pet blend ratio on ring spun yarn tenacity. Periodicals of Engineering and Natural Sciences. 2017;**5**:176-180. DOI: 10.21533/pen.v5i2.110

[33] BS EN ISO 139:2005+A1:2011Textiles. Standard atmospheres for conditioning and testing

[34] BS EN ISO 2062:2009 Textiles. Yarns from packages. Determination of single-end breaking force and elongation at break using constant rate of extension (CRE) tester

[35] ISO 16549:2004Textiles. Unevenness of textile strands – Capacitance method

[36] Nicolic M, Stjepanovic Z, Lesjak F, Stritof A. Compact spinning for improved quality of ring-spun yarns. Fibres & Textiles in Eastern Europe. 2003;**11**:30-35

[37] Jackowski T, Cyniak D, Czekalski J. Compact cotton yarn. Fibres & Textiles in Eastern Europe. 2004;**12**:22-26

[38] Altas S, Kadoğlu H. Comparison of conventional ring, mechanical compact pneumatic compact yarn spinning systems. Journal of Engineered Fibers and Fabrics. 2012;**7**:87-100

[39] Dash JR, Ishtiaque SM, Alagirusamy R. Properties and processibility of compact yarns. Indian Journal of Fibre & Textile Research. 2002;**27**:362-368

[40] Yilmaz D, Usal MR. A comparison of compact-jet, compact, and conventional ring-spun yarns. Textile Research Journal. 2011;**81**:459-470. DOI: 10.1177/0040517510385174

[41] Mamun RA, Repon MR, Jalil MA, Uddin AJ. Comparative study on card yarn properties produced from conventional ring and compact spinning. Universal Journal of Engineering Science. 2017;**5**:5-10. DOI: 10.13189/ujes.2017.050102

Chapter 3

Flame-Retardant Unsaturated Polyester Resins: An Overview of Past and Recent Developments

Ewa Kicko-Walczak and Grażyna Rymarz

Additional information is available at the end of the chapter

http://dx.doi.org/10.5772/intechopen.72536

Abstract

The chapter presents a positive effect of reduced flammability of thermoset resins, thanks to the use of halogen-free flame retardants (FRs) and multi-ingredient modifiers that combine conventional phosphorous/nitrogen additives interacting with nanofillers. The impact of these inhibitors on the level of flammability of polyester polymer compositions was defined by determining the value of limiting oxygen index (LOI), thermogravimetric analysis and cone calorimeter (CC) analysis of thermal destruction processes. Morphology of composites was assessed by scanning electron microscope and analysis of actual SEM micrographic images. The analysis of thermal decomposition of the materials under examination confirmed flammability-reducing properties of the hybrid flame retardants used, and a synergy effect was observed between conventional modifiers and nanofillers, specifically expandable graphite and synthetic precipitated silica. The multi-ingredient composition of flame retardants turned out to make significant progress in achieving the desired level of flammability in polyester resins. Nanocomposites are a new class of polymer systems.

Keywords: flame retardancy, halogen-free antipyrine, nanofillers, nanocomposites, flame retardants, polyester resins

1. Introduction

Unsaturated polyester resins are the most commonly used thermoset resins in the world. More than 2 million tonnes of unsaturated polyester resins are utilized globally for the manufacture of wide assortment of product, including pipe, tanks, sanitary ware and high-performance components for the building, marine, electric and automotive industries. High thermal and heat stability, low shrinkage and excellent mechanical properties are typical for their polymers. Those applications in addition to the mechanical and electrical properties also require good flame retardants of the materials.

The contemporary chemistry of polymers and polymeric materials faces extraordinarily high requirements as regards broadly understood ecology of production processes and product properties, including reduced flammability and lower emission of toxic compounds generated as a result of thermal destruction of polymer materials.

This applies mainly to structural polymer materials that we encounter in daily life (construction, transport, household accessories and sports articles).

One of the commonly and broadly used groups of structural polymers comprises specifically unsaturated polyester and epoxy resins. Therefore, the problem of suppressing flammability of polyester and epoxy resins is currently one of the most serious and at the same time most difficult challenges for the state-of-the-art technologies for the production of these materials.

While burning, unsaturated polyester resins and polyester glass laminates made with these resins give off significant quantities of thick black smoke. This phenomenon is accompanied by the emission of toxic gases (HCI, HBr, CO), especially if flame-retardant halogen polyester resins have been used. Rescue operations become harder or even impossible.

Therefore, it is important that polyester resins are obtained that do not sustain the burning process, characterized at the same time by an appropriate level of flame retardancy and reduced emission of smoke.

Reducing flammability of thermoset materials by the addition of halogen compounds or inclusion in the structure of these polymers of a chemically bound halogen has been discontinued on a successive basis for ecological reasons. These compounds, even though they are very effective in decreasing flammability of thermoset materials and show favorable synergy effect, mainly in the halogen-antimony system, are characterized by the emission of toxic products in combustion, mainly hydrogen chloride and hydrogen bromide. Their strong corrosive and irritant action, resulting from the use of aromatic bromine compounds, is a reason for limiting the use of this method and searching for alternative solutions. Undertaken activities refer to official drafts, laws and legal recommendations. In the territory of the European Union, the REACH Regulation (with supplementary appendices) of the European Parliament and the Council has been in force since 2007, as the regulation for managing chemical policy in EU states [1]. The main objective of REACH is to control introduction of and transactions in already marketed chemical compounds in terms of safety of human health and life and natural environment. One of the major elements of legislative regulations (EU REACH regulation) is to assess especially controversial substances and to gradually withdraw them from the EU markets. In addition to REACH, EU states are also bound by the RoHS Directive, which includes a ban on using halogen derivatives in the production of electronic and electric equipment. Most conventional flame retardants like halogenated organics have been limited in their utility due to peoples' health and environmental concerns, and other fire retardants like aluminum hydrates (ATH) are expensive and need to be used in large quantities which the thermoset polymers lose its original mechanical properties. The application of ATH or magnesium hydroxide requires a very high loading of the filler within the polymer matrix. Loading levels of more than 60 wt.% are necessary to achieve a suitable flame retardancy. The disadvantages of these loading levels are high density and lack of flexibility of the final products, as well as low strength properties and problem in extrusion/compounding technical process.

These legislative initiatives are a part of justified attempts and a necessity of searching for new, equally effective methods of reducing flammability of thermoset materials. Currently, reducing flammability of these polymers focuses on the use of halogen-free inorganic compounds, which, as additive modifiers, do not form chemical bonds with a polymer. From an industrial practice perspective, it is important that the desired flammability level is achieved at the lowest possible concentration of a flame retardant (FR) [2]. Therefore, the current tendency among academic institutions is to focus on the use of multi-ingredient halogen-free FR systems, taking advantage of the synergy effect of their action [3]. Synergy effects should be considered one of the most important phenomena related to the issues involving retardation of polymer flammability.

Recently, increased interest in to the use of additions in the form of nanocompounds as modifiers of multiple properties of polymers, including their thermal stability and fire resistance. Large interfacial surface and nanometric dimensions of modifier particles differentiate nanocomposites from traditional composites in a significant way. Nanocomposites avoid the disadvantages of these conventional flame-retardant systems. Therefore, the assumption that nanomaterials formed this way make it possible to achieve significant progress and benefits for human kind and natural environment due to their distinctive properties compared to the properties demonstrated by macrometric size particles has also rolled out to the issue of effective reduction in flammability of nanocomposite materials [5–7, 14].

To produce polymer nanocomposites, including polyester ones, natural layered aluminum silicates are usually used, such as smectites, kaolinites, halloysite nanotubes, including mainly montmorillonite (MMT), or modified compounds—by ionic exchange of MMT-based clay with quaternary ammonium salts. Layered silicates dispersed in a polymer matrix are the most popular form of hybrid nanocomposites [16, 17]. Commonly used MMT-modifying ammonium salts, which contain long alkyl carbon chains in their structure, make aluminum silicates acquire organophilic properties, improve their affinity to a polymer matrix and thus reinforce composites' thermal stability which is determined by the length, number and saturation degree of alkyl chains. Nanocomposites modified this way demonstrate better thermal resistance, lower flammability and advantageous strength properties [13]. This concerns, however, mainly nanocomposites obtained with the use of polyolefines, polyamide, polystyrene [9–11], PVC [12] and other thermoplastics and elastomers, such as rubbers—SBR, BR and HNBR [4, 5, 8].

The results of experiments conducted in recent years on thermoset materials with modifications by mere MMT or organophilic MMT (oMMT) have confirmed a positive impact of this modification on strength properties, including mainly a significant increase in Young modulus of the products, but at the same time it has turned out that their flammability reducing effect is less favorable [18, 19].

The analysis confirms these observations and suggests a reason for this phenomenon. An effective multi-ingredient system has been proposed, which combines the joint effect of conventional flame retardants and selected nanocompounds, including expandable graphite and synthetic precipitated nanosilica [20]. The silica used in the experiments is a synthetic silicon oxide obtained by acidifying sodium silicate with sulfuric acid. In order to improve compatibility of silica and polymer matrix, acidification process was modified to obtain a suspension of precipitated silica, and then, this suspension sustained separation and drying processes. The resultant silica in the form of spheroid pearls was pulverized or subjected to wet granulation. It should be noted that precipitated silica contains a number of silanol groups, which do have a positive impact on interaction with a polymer, but they also lead to reinforcement of hydrophilic properties. Therefore, in the course of analysis, modification of silica with pro-adhesive compounds was suggested, such as silane and borate titaniate compounds. Control of these parameters, combined with the control of pH, temperature, reaction duration, and molar ratio of reagents, makes it possible to obtain a portfolio of products from the group of nanosilica modifiers. In this analysis, nanosilica of the optimum set of properties was used, said properties being a sign of a possibility to obtain nanocomposites characterized by favorable thermal stability and lower flammability.

In order to reinforce flame retardation effect by the use of mere nanocompounds, polyphosphoric polymers were mainly used, intumescent flame retardants (IFR) in the form of melamine polyphosphate, ammonium polyphosphate and melamine borate, showing a proven effectiveness in inhibiting the flame. In this process, benefits of phosphoric-nitrogen flame retardants were realized, showing a synergy effect in reducing heat release rate (HRR) and positive effects in the reduction of flammability of structural compounds while using boron compounds [15, 21, 22].

An important part of the work, consistent with the authors' intent, refers to taking advantage of the benefits of using hybrid flame retardants, combining phosphorous-derived modifiers of thermoset materials' combustion process with selected modifiers when compared to the effect of reducing flammability level by using a single ingredient of the system only [23, 24]. In the course of experiments, optimum recipes and technological parameters for obtaining composite thermoset resins (unsaturated polyester resins and epoxy resins) with the use of multi-ingredient flame retardants were investigated. It was suggested which mechanism to use and how a deliberately chosen selection of multi-ingredient FR modifiers can have a positive impact on flame inhibition and lower smoke emissions.

2. Experiments

2.1. Materials and raw materials

Polimal 103A—structural orthophthalic unsaturated polyester resin, produced by Zakłady Chemiczne "Organika-Sarzyna" CIECH S.A. Viscosity in the range of 300–500 mPa.s, styrene content of 35% (UP_1) or equivalent orthophthalic polyester resin, DION 9500 Series produced by Reichhold, Inc. Company (UP_2).

Hardening system: two parts by weight of methyl-keton hydrogen peroxide Luperox K-1 and 0.4 parts by weight of cobalt naphthenate (1% Co).

Zinc borate: $ZnO\ B_2O_3{\cdot}2H_2O$ (ZnB)—commercial name of the product Flamtard Z-10 is product of Waardals Comp.

Zinc stannate/hydroxy zinc stannate: $ZnSnO_3/ZnSn(0H)_6$ (Zn/Sn)—commercial name of product Flamtard S/Flamtard HB1 was produced by Alcan Chemical Europe.

Melamine polyphosphate(MPP): crystalline form of P_2O_5 content = 32–33% and N = 41–42%, average grain size of 2–3 μm, was manufactured by Zakłady Chemiczne" Alwernia"S.A.

Ammonium polyphosphate (APP) was manufactured by Clariant GmbH, commercial name of the product: AP 412.

Melamine borate (MB): experimental product of the Institute of New Chemical Synthesis is branch of the Institute of Inorganic Chemistry in Gliwice.

Montmorillonite (MMT) and MMT modified with quaternary ammonium salts (oMMT) was manufactured by Southern Clay Products, commercial name of the product: Nanofil 5.

Expandable graphite: EG 096 (EG) was manufactured by Synograf S.A.

Synthetic precipitated silica ($nSiO_2$): experimental product of the Institute of New Chemical Synthesis is branch of the Institute of Inorganic Chemistry in Gliwice.

2.2. Method of obtaining flame-retardant polyester compositions

For technical reasons, it was decided to introduce all flame retardants into ready-made polyester resins before the addition of hardening agents. Thanks to this procedure, it was possible to modify different types of resins in practice, at both resin manufacturers and processing entities. Additions were stirred in using a low rotation speed mixer per approx. 5 min. After first stirring with a low speed rotation mixer, the process was continued in an ultrasonic mixer provided with a turbine mixer (approx. 4500 rev/min). Resultant compositions were additionally ground in a high rotation speed homogenizer Mixman Comp., especially if nanocompounds were used in a composition. It was a guarantee of obtaining a stable and homogenous composition of resins with flame retardants after approx. 80–90 min of stirring.

2.3. Testing methodology

2.3.1. Limiting oxygen index

Following industrial practice, measurements of limiting oxygen index (LOI) were used as a preliminary indication of relative flammability, said index representing mainly the level of flammability of volatile products of polymer decomposition. Measurement of LOI carried out to the standard PN-EN ISO 4589–2:1999/A1:2006 [25].

2.3.2. Thermogravimetric analysis

Thermogravimetric analysis was performed using a thermogravimetric analyzer TGA/SDTA 851 device by Metter Toledo. Samples of 10 g were placed in platinum melting pots with the capacity of 150 g. Heating up progressed in defined cycles, namely within the range of 25–800°C in air or nitrogen atmosphere and with a heating rate of 20°C/min.

2.3.3. Cone calorimeter tests

In the presented work, tests were performed using a British cone calorimeter FTT Dual Analysis Cone Calorimeter. Using computer analysis, a number of flammability properties of materials were defined in line according to ISO 5660:2002 [26], part 1 "Reaction to fire tests – head release rate, smoke production and mass loss rate" using an external heat stream of a density of 25 or 50 kW/m^2 and air flow rate of 24 l/s. Type of burning reaction initiation-ignition was with a horizontal orientation of samples relative to a radiator. A spark fuse was used to ignite gases. Frequency of data compilation was 5 s for each.

Tests were performed on previously air-conditioned samples, 100 × 100 × 10 mm, having constant relative moisture and constant mass.

2.3.4. SEM tests

SEM tests were performed using a scanning electron microscope Hitachi SU8010. Tests were performed on specimen sections of sample test blocks and hardened resin composites after their modification.

SEM tests were also used to analyze the surface of tested composites after charring as a result of cone calorimeter tests.

3. Results overview

3.1. Halogenated unsaturated polyester resins

Halogenated unsaturated polyester resins are produced mainly on the basis of HET acid (hexachloroendomethylenetetrahydrophthalic acid), obtained by reacting maleic anhydride with hexachlorocyclopentadiene. Polyesters based on this acid do still have industrial importance and continue to be used in polyester resins produced by European and American companies, like Ashland Groups HETRON among others.

The production of HET acid-based polyester resins was also launched in Poland at the beginning of 1960s. The trade name of the product was Polimal 162, and it was widely used for the construction of flame-retardant rescue boats. The production of a new type of an unsaturated polyester resin with HET acid, characterized by better chemical resistance, was restored in the Chemical Plant "Organika-Sarzyna" in 1997.

Hooker Chemical Co. launched the process of chlorinating phthalic anhydride to tetrachlorophthalic acid anhydride, and Michigan Chemical was brominated phthalic anhydride to tetrabromophthalic acid anhydride. Tetrahalogenphthalic compounds were a basis for several flame-retardant unsaturated polyester resins launched at the turn of 1960s and 1970s. In the same period, attempts were made to favorably introduce bromine into a polyester chain by the use of brominated dioles (e.g., dibromoneopentyl glycol) or direct use of elementary bromine. What should be also mentioned is the use of tetrabromobisphenol A, which became a basis for the launch of several types of unsaturated polyester resins, type Derakane, characterized by good resistance to thermal shock and particularly favorable resistance to corrosion. Another method of introducing bromine atoms into a resin structure, mainly in the case of vinyl-ester resins, is the use of brominated cross-linking agent, such as dibromostyrene.

However, one should note that halogenated unsaturated polyesters can be obtained not only by polycondensation of dicarboxylic acids and (maleic, phthalic) acid anhydrides with glycols, but also by copolymerization of acid anhydrides with epoxy compounds, for example, 1,1,1-trichloro-2-3-epoxypropane or epichlorohydrin. Thus, chlorinated polyester resins with reduced flammability was obtained. An additional advantage of the technology is that it is waste free. One of the first industrial products obtained by catalytic addition of epichlorohydrin to dicarboxylic acid anhydrides, which we describe in detail in the next section of this resume, was Polimal 160 produced by the Chemical Plant "Organika-Sarzyna"-CIECH.

3.2. Halogen-free flame-retardant modifiers in polyester resins.

3.2.1. Modification of resins by using tin-zinc compounds

In 2003, we started experiments on the reduction of flammability of polyester resins using tin and zinc compounds. For experiments, we used oxides and hydroxides of these metals, which constituted a new group of halogen-free flame retarders relative to thermosetting resins. The results obtained by experiments make the flame retarding efficiency of these compounds more specific compared to the existing commonly used flame and smoke suppressant, that is, antimony trioxide (Sb_2O_3).

The experiments with the new compounds led to the following conclusions:

1. Compared to the experiments testing thermal degradation of unsaturated polyester resins modified with $ZnSn(OH)_6$, we concluded that thermal decomposition proceeded in a complex way consisting in the precipitation of a series of low-molecular products coming from the decomposition of structures containing phthalic anhydride, and then, at higher temperatures, styrene, benzoic acid and other aromas. The analysis of the decomposition process for the samples modified with Sn/Zn indicates that the intensity of CO_2 emission was much reduced, and at the further decomposition stage, trace amounts of water and HBr occurred. Based on the experiment results, we proposed a thermal decomposition mechanism for modified unsaturated polyester resins of reduced flammability.

2. Following the actual kinetic analysis of the thermal decomposition process, we found that an unsaturated polyester resin decomposed in three stages. The first one was characterized by the activation energy of around 130 kJ/mol, comparable to depolymerization energy of polymerized polystyrene. The second stage, with activation energy of 150–170 kJ/mol, could be attributed to the decomposition of a cross-linked polyester network, wherein the third stage ($E > 200$ kJ/mol) was random macro-chain decomposition, leading to the formation of low molecular decomposition products. For a sample modified with $ZnSn(OH)_6$, we concluded that introduction of an addition leads to an increase in (apparent) activation energy for the area of $\alpha > 0.8$. The kinetic analysis by nonlinear regression showed that for a nonmodified polyester resin, the best statistic distribution (F-test) corresponded to the Avrami-Erofeev (nuclei growth) model, and for polyester compositions modified with $ZnSn(OH)_6$, the best distribution was for the n-th order autocatalytic reaction.

3. Stabilizing impact of $ZnSn(OH)_6$ can be explained by the formation of a spheroid surface barrier blocking or inhibiting the heat flow from the decomposition zone to the substrate and by reacting tin compounds with radical decomposition products of a high-molecular compound.

4. When it comes to the choice of the most effective examination of resins reflecting their behavior in a fire in the most reliable way, we chose the experiments performed with the use of a cone calorimeter according to ISO 5660: 2002. The analysis of the combustion process of polyester compositions conducted with the cone calorimeter method, being the state-of-the-art and primary method of examining polymers' flammability, let me conclude that the presence of Sn/Zn flame retarder led to the reduction in the heat release rate and total heat release, significant reduction in smoke emission and longer time required for the ignition of the composition.

5. The results are illustrated by the following curves and diagrams presenting the selected figures (**Figures 1–4**) of the analysis of composition processes for polyester composites modified with antiflammable Sn/Zn in juxtaposition with analogically prepared samples of nonmodified polyesters. The experiments were done at heat output of 25–50 kW/m^2.

Analysis of the thermoanalytical experiments and combustion process was a basis for the proposal of the $ZnSn(OH)_6$ thermal stability mechanism for unsaturated polyester resins:

UPR/UP $\rightarrow$ phthalic anhydride, styrene, aromas + $R\text{-}CH_2\text{-}Br$,

where R – hydrocarbon chain,

$R\text{-}CH_2\text{-}Br + [ZnSn(OH)_6 \rightarrow ZnSnO_3 + H_2O] \rightarrow Zn_2SnO_6 + SnBr\chi$ + hydrocarbons,

where: $\chi = 2$ or 4.

$SnBr_2 + H_2O \rightarrow (SnO, SnO_2) + HBr$.

Resultant hydrogen bromide acts as a radical "catcher."

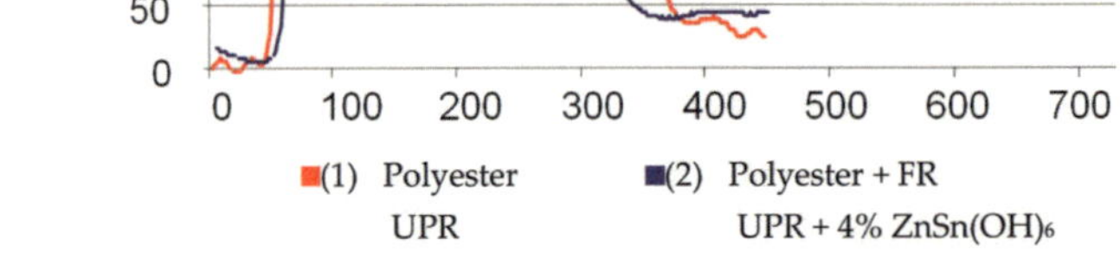

Figure 1. Heat release rate (HRR) at heat exposure of 50 kW/m^2.

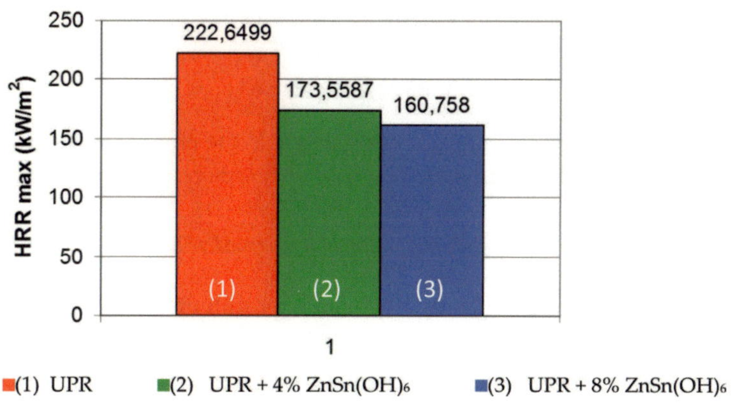

Figure 2. Maximum heat release rate (HRR max).

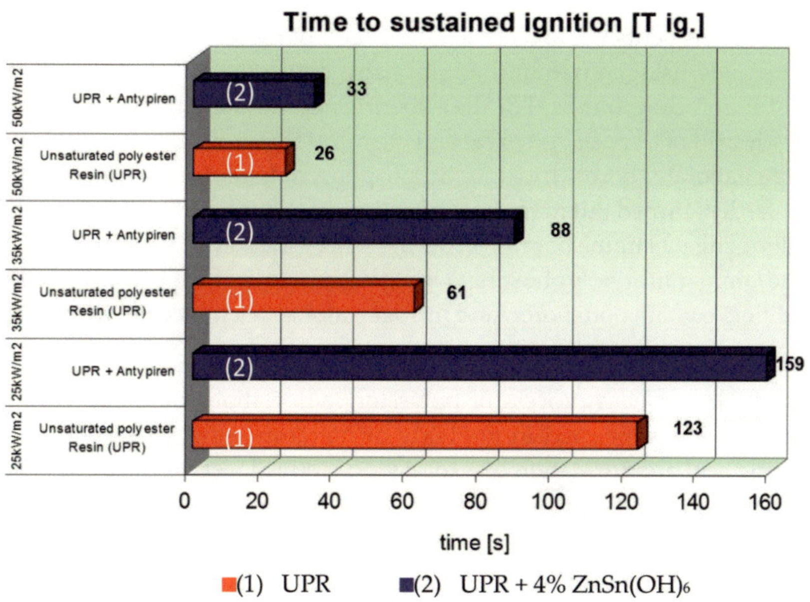

Figure 3. Time to sustained ignition (T ig) at different heat exposures of 25–50 kW/m^2.

It can also be assumed that H, OH and RO radicals formed as a result of unsaturated polyester resin bond decomposition can take part in the reactions with SnO and SnOH:

$$SnO + H \rightarrow SnOH. \tag{1}$$

$$SnOH + H \rightarrow SnO + H_2O \tag{2}$$

$$SnOH + OH \rightarrow SnO + H_2O \tag{3}$$

$$SnOH + RO \rightarrow SnO + ROH \tag{4}$$

The aforesaid reactions form the so-called catalytic cycle.

The aforesaid findings confirm the suggestions related to the effective action of $ZnSn(OH)_6$ as a flame-retardant agent stabilizing unsaturated polyester resins in the conditions of a fire risk.

3.2.2. Use of boron compounds as flame retarders for unsaturated polyester resins

A logical addition to the abovementioned scope of experiments focusing on the reduction of flammability of polyester resins was my concept to use boron compounds and boron salts as FR modifiers. We took an assumption that instead of the known antimony oxides cooperating with a halogen embedded in a polyester chain, one of zinc borate forms could be used, namely $ZnO{\cdot}B_2O_3{\cdot}2H_2O$. It turned out to be an effective flame retarder for the majority of halogen unsaturated polyester resins, specifically for those, for which chlorine or bromine atoms were bonded with aliphatic or cycloaliphatic and not aromatic fragments of a molecule The results of these experiments made us check the effects of modifying the selected unsaturated resins with the new halogen-free and additionally physiologically neutral flame retarders in the form of boron compounds. For the experiments, we used the following compounds: anhydrous boric acid of a special physical and chemical structure, zinc borate, calcium borate, ammonium pentaborate, boron trioxide, boron phosphate and melamine borate. The new boric flame retarders turned out to be more effective than zinc borate used so far. We achieved the best results using melamine borate. As for the other FR additions, ammonium pentaborate and boric acid/ammonium polyphosphate system are worth mentioning. As all of the most effective modifiers contain both boron and nitrogen atoms in a molecule, we could assume the synergy action of these elements.

The analysis of the pyrolysis process of polyester compositions modified with boron compounds confirmed—as suggested in the body of literature—the generation of a high temperature of boric acid and formation of a foamy protective layer on the composition surface, limiting or even stopping the process of a product's burning. While using melamine borate,

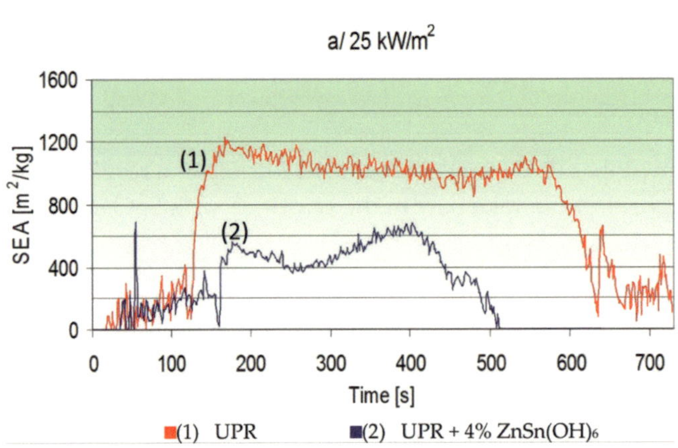

Figure 4. Smoke extinction area at radiation of 25 kW/m^2.

we also found the smoke suppressing and char incandescence reducing action after the extinction of the flame burning. We noticed that this modifier improved the "enclosure" of the surface of the sample burnt by the formation of a d nse protective film.

3.2.3. *Flame-retardant polyester resins by using hybrid modifiers*

Initial assessment of an impact of a compilation of halogen-free flame retardants of flammability level of polyester compositions, performed as LOI evaluation, pointed to an especially favorable impact of melamine polyphosphate used in combination with expandable graphite/precipitated silica. A relatively small quantity of a combination of these flame retardants used in a proportion of 15 parts by weight to 5 parts by weight, respectively, led to an increase in LOI by 20–35% for samples of resins without and with reinforcement of a glass-reinforced polyester resin fiber compared to nonmodified resin only (**Table 1**) The polyester glass laminates GRP were obtained with the discussed composition as a matrix was characterized by LOI > 32–35%.

MMT modifiers showed a slightly worse flammability reduction impact, which was commented upon in the introductory part of the publication.

Thermogravimetric analysis of selected polyester compositions was performed in dynamic conditions at a heating rate of 20 K/min in air or nitric conditions—(**Figure 5**). Analysis of thermogravimetric curves (TGA) of the process of sample decomposition confirmed a different course of thermal decomposition of a nonmodified sample compared to compositions tested analogically and containing multi-ingredient flame retardants. The sample in which FR ware made mass loss is much smaller at the same temperature as compared to reference sample. This is

Resins	**Antipyrines**		**Oxygen index (LOI). % PN-EN ISO 4589-2:1999/A1:2006**	
	Conventional FR + Nanofillers	**Content, weight %**	**Polyester composites**	**Polyester GRP laminates**
Polimal 103A–UP_1	-	-	19.7	-
DION 9500-UP_2	-	-	21.2	22.8
UP_1	MPP	15	2.5	33.1
UP_1	MPP + oMMT	15 + 5	28.1	34.1
UP_1	MPP+ EG	15 + 5	27.3	35.8
UP_1	MPP + n SiO_2	15 + 5	28.7	-
UP_2	MPP	15	26.8	3.7
UP_2	MPP + oMMT	15 + 5	27.7	3.3
UP_2	APP + oMMT	15 + 5	27.2	-
UP_2	MB + oMMT	15 + 5	27.5	-

Table 1. Determination of limited oxygen index according to PN-EN ISO 4589-2:1999/A1:2006.

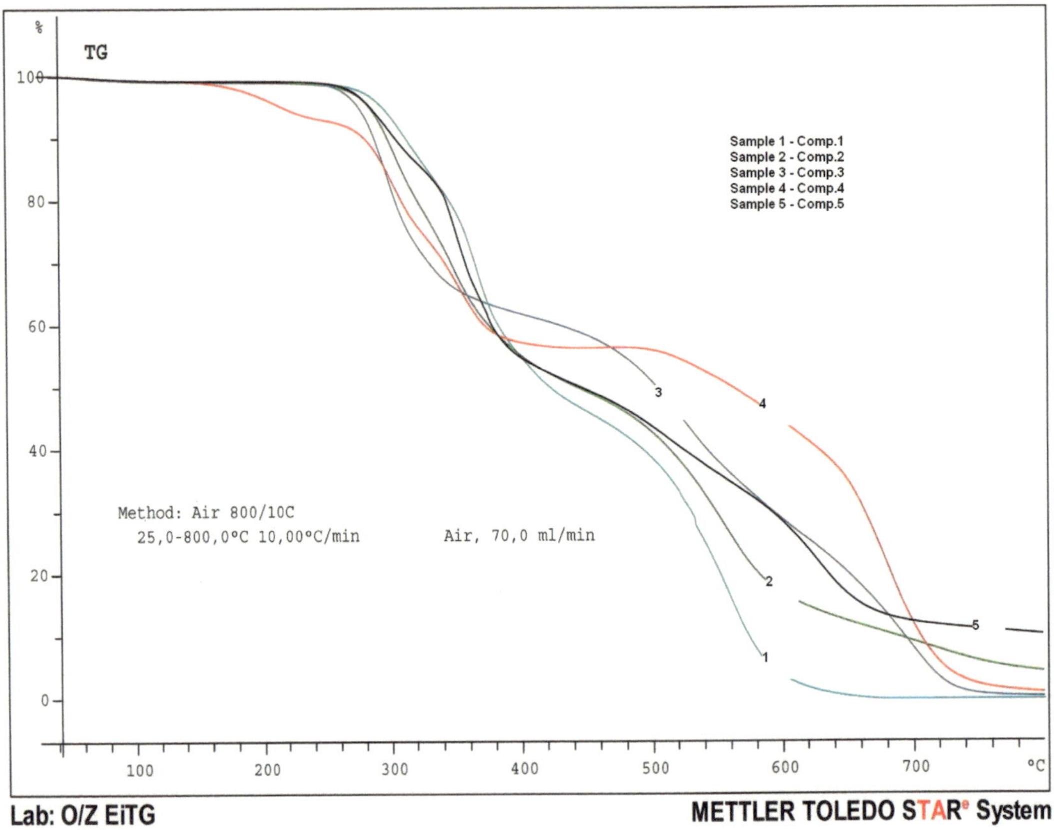

Figure 5. Thermogravimetric analysis of selected polyester compositions: 1-UP, 2-UP + 10MPP + 5oMMT, 3-UP + 10MPP + 5nSiO_2, 4-UP + 10MPP + 5EG, 5-UP + 15MPP + G. time, s,UP—unsaturated polyester resin, MPP—melamine polyphosphate, MMT—organophilic montmorillonite, nSiO_2—nanosilica, EG—expandable graphite.

the proof of the retardant action of these N/F with nanofiller system. As one can see, thermal decomposition of a non-modified polyester resin starts at around 300°C and proceeds in a single stage up to a temperature of around 430°C. *maximum* decomposition rate takes place at a temperature of around 350°C as determined on the basis of the DTG curve. Samples which contain a flame retardant, especially MPP and expandable graphite or precipitated nanosilica, undergo basic decomposition at around 420°C. Introduction to the system of multi-ingredient, especially selected flame retardants led to an increase in the quantity of solid decomposition leftovers.

The unmodified UP and the selected composites of UP with the PPM/nanomodified were investigated by means of TG analysis. For all samples, the process of thermal decomposition begins above 200°C, the differences in weight loss from samples appear and they become more distinct with an increasing temperature. The nonmodified UP loses 98% of its weight at 450st.C due to almost complete combustion process in nitrogen atmosphere. Referring to expectations, UP with FR modifiers (PPM 15 wt.% with EG 5 wt.%), has lower weight loss (85%) at the same temperature. This is partly due to high thermal stability of EG component, which loses only about 25% of its weight at 450st.C. For UP containing the compounding of PPM 15 wt.% and 5 wt.% oMMT, a slightly higher weight loss was observed. The same weight loss at 450st.C was obtained for UP modified with the compounds of PPM/EG used in the amount of 25 weight parts per 100 parts of UP. All the modified UP with PPM/EG additives yields about 10 wt.% of char residue at 600st.C, which also may contribute to their FR action shown on the basis of flammability analysis.

The cone calorimeter (CC) measuring principle is that of oxygen consumption. This states that there is a constant relationship between the mass of oxygen consumed from the air and the amount of heat released during the polymer degradation.

The cone calorimeter test was followed up by HRR curves illustrating changes in heat release rate as a function of burning time. Heat release is shown as power per unit of exposed test surface. HRR curves for a polyester without and with fireproofing in the function of heat flux 50 kW/m^2 are shown in **Figure 6**. Analysis of HRR and MLR (**Figure 7**) curve shapes for compositions with flame retardants makes it possible to conclude that after the actual maximum increase in HRR, heat release rate decreases in a significantly shorter time compared to an analogically tested sample of a nonmodified polyester sample, and there is no second HRR peak. It should be contributed to an isolating effect of flame retardants, thanks to which heat release rate gets stable over time depending on radiation power at constant.

These results tend to demonstrate that the improvements in flame retardancy for nanocomposites do not occur in the process in the gas phase but rather by a change of the combustion act in the condensed phase (**Figure 8**).

Analysis of SEM micrograph presented in **Figure 9** makes it possible to conclude that a multi-ingredient flame retardant was distributed evenly over a polymer matrix, and additions grinding level is sufficient to have a homogenous system without a tendency to agglomerate.

In **Figure 9a**, showing this test, MPP in combination with MMT was used in order to additionally confirm homogenous morphology of compositions obtained this way.

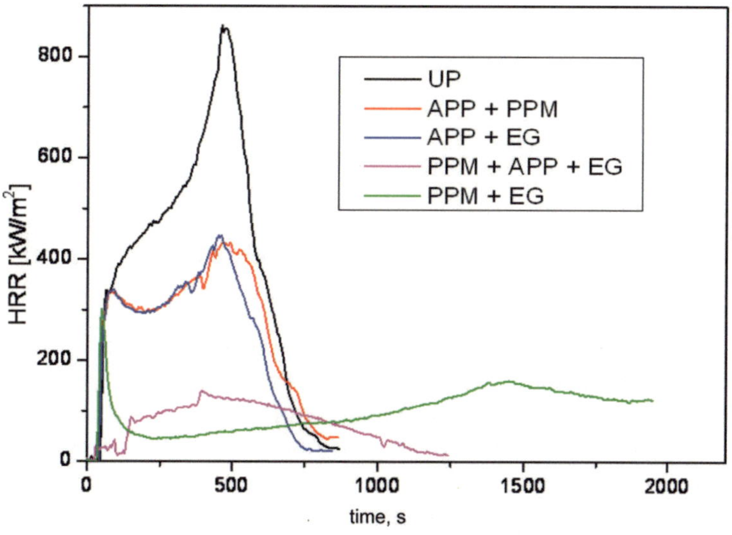

Figure 6. Course of heat release rate (HRR) versus combustion time (unsaturated polyester resins modified flame retardants).

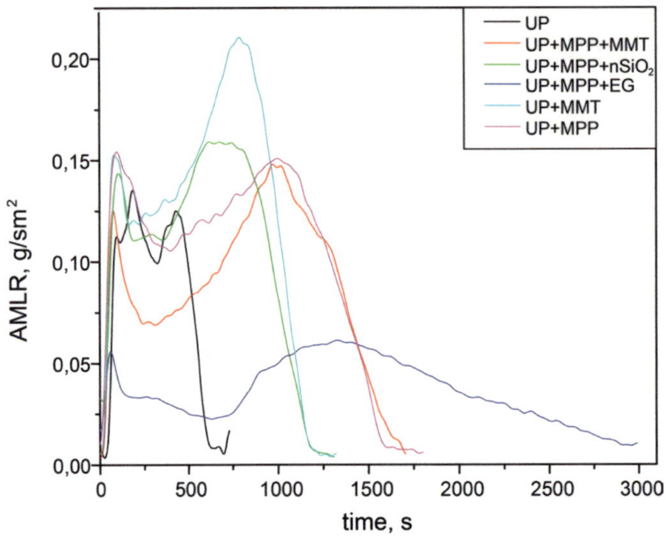

Figure 7. Course of mass loss rate (MLR) versus combustion time (UP resins).

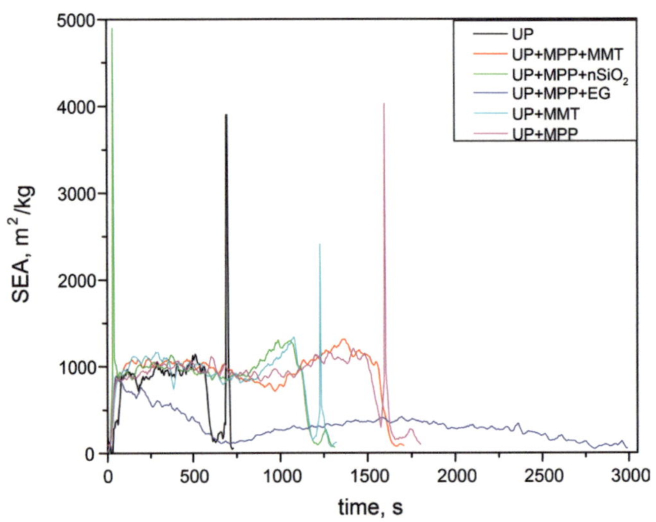

Figure 8. Course of specific extinction area (SEA) versus combustion time (UP resins).

SEM test performed for the surfaces of polyester compositions tested before and after burning by a cone calorimeter method confirmed intumescent impact of a phosphorous-nitrogen-based modifier, namely melamine polyphosphate MPP. In **Figure 9b**, showing this test, MPP in combination with EG was used in order to additionally confirm homogenous morphology of compositions obtained this way.

Synergy effect of expandable graphene, nanosilica and MMT with conventional flame retardants and their positive impact on thermal resistance and polymer burning profile were confirmed. Our tests suggested that there are at least two processes which reduce flammability

a)

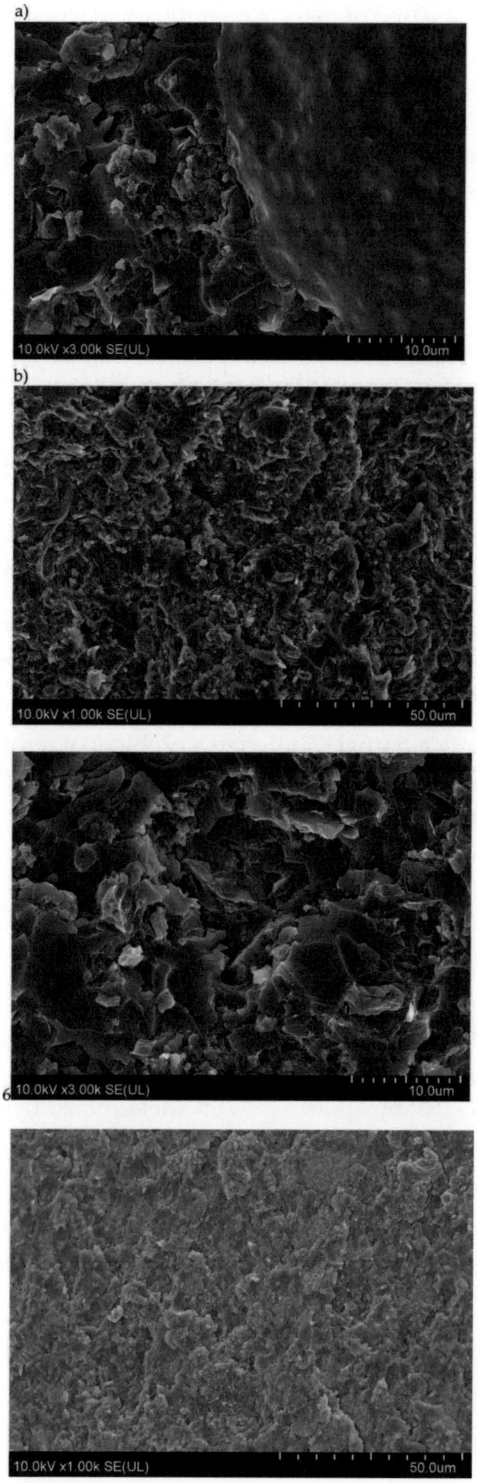

Figure 9. SEM micrographs (a,b) – Cross section of polyester resins modified with inorganic–organic fire retardants: a) SEM image of UP +15% MPP+ 5% o-MMT, b) SEM image of UP + 15% MPP + 5% EG.

level of polymers tested. In the first of them, a nanofiller had a positive impact on coke formation, and in the other one, non-flammable organic material influenced nanosilica /MMT dispersion, and in consequence, barrier properties reduce gas diffusion. Synergy effect was also observed while using traditional flame retardants and expandable graphite or while modifying selected melamine salts and this graphite structure which is active mainly in the solid phase, reducing toxic fume emissions in an effective way.

4. Conclusions

1. Halogen-free modifiers give possibility to obtain ecological polymer composition and a new class of flame-retardant thermoset polymer systems.
2. Commercially available flame retardants such as ATH or halogen-containing compounds are effective; however, the mechanical properties and processing of those polymers are often negatively influenced by the large quantities of halogen FR additive that are needed. The addition of those FR often increases the production of soot and carbon monoxide.
3. Nanocomposites have many advantages over the traditional flame retardants. Processing of nanocomposites is straightforward, and as the nanocomposites, no additional halogen is considered as an environmentally friendly alternative.
4. Nanocompounds are only a partial solution to a problem of reducing flammability of thermoset polymers. oMMT clay impact, mainly due to significant difficulties in its proper dispersion in a polymer matrix, is not fully convincing. Large specific surface of a nanofiller and a tendency to form a carbonized barrier layer should indeed promote thermal stability and reduction in a composite flammability level, but the catalytic effect of intercalation elements (flammable quaternary ammonium salts) in oMMT may lead to increased thermal degradation of thermoset polymers.
5. An effective solution was to use phosphorous-ammonium-based modifiers as conventional modifiers in a compilation with nanocompounds, mainly expandable graphite and precipitate nanosilica. Multi-ingredient flame and smoke inhibitors showed a positive impact on significant reduction in HRR by approx. 38–70% (determined by a CC test), increase in LOI by around 20–35%, and favorable profile of thermogravimetric curves (TGA) for modified polymers.
6. Micrograph images of nanocomposites performed by SEM confirmed good distribution of multi-ingredient systems in a duroplastic matrix without a tendency to agglomerate or sediment. Multi-stage technique of modifier intercalation with an ultrasonificator turned out to be an effective method of obtaining homogenous nanocomposites with the share of tested resins.
7. The concept of using multi-ingredient flame and smoke retardants made is possible to reduce the quantity of flame retardants used for obtaining the desired flammability level in products. This fact contributes to slightly negative influence of the modifications on

physical and chemical properties and strength parameters of polymers after setting. It is a significant step ahead for the techniques in the processing of thermoset polymers, and nanocomposites have many practical advantages over traditional flame retardants.

8. A multi-ingredient combination of flame retardants turned out to make significant progress in achieving a desired flammability level in thermoset resins. It should be assumed that the presence of conventional flame retardants led to a synergy effect promoting faster formation of a carburized protection layer hindering oxygen flow-through and emission of low-molecule compounds resulting from the process of polymer thermal destruction.

Author details

Ewa Kicko-Walczak* and Grażyna Rymarz

*Address all correspondence to: ewa.kicko-walczak@impib.pl

Institute for Engineering of Polymer Materials and Dyes in Toruń, Branch of Paints and Plastics, Gliwice, Poland

References

[1] Directive 2002/95/EC of the European Parliament and of the Council of 27 January 2003 on the restriction of the use of certain hazardous substances in electrical and electronic equipment and Directive 1907/2006/EC of the European Parliament and of the Council of 18 December 2006 concering the Registration, Evaluation, Authorisation and Restriction of Chemicals

[2] Betts KS. New thinking on flame retardants. Environmental Health Perspectives. 2008; **116**:A210-A213

[3] Kicko-Walczak E. Studies on the mechanism of thermal decomposition of unsaturated polyester resins with reduced flammability. Polymers and Polymer Composites. 2004;**12**:127

[4] Saritha A, Joseph K, Thomas S, Mutaleekrishman R. Chlorobutyl rubber nanocomposites as effective gas and VOC barrier materials. Composites: Part A. 2012;**43**:864

[5] Kader MA, Kim K, Lee Y-S, Nah C. Preparation and properties of nitryle rubber/montmorillonite nanaocomposite via latex blending. Journal of Materials Science. 2006;**41**:7341

[6] Alongi J, Brancatelli B, Rosace G. Thermal properties and combustion behaviour of POSS-bohemie-finished cotton fabrics. Journal of Applied Polymer Science. 2012;**132**:426

[7] Huang G, Zhuo A, Wang L, Wang X. Peraration and flammability properties of intumescent flame retardant-functionalized layered double hydroxides/polymethyl methacrylate nanocomposites. Materials Chemistry and Physics. 2011;**130**:714

[8] Choundhury A, Bhowmick AK, Ong Ch. Effect of different nanoparticles on thermal, mechanical and dynamic mechanical properties of hydrogenated nitrile butadiene rubber nanocomposites. Journal of Applied Polymer Science. 2010;**116**:1428

[9] Lao SC et al. Flame–retardant polyamide 11 and 12 nanocomposites: thermal and flammability properties. Journal of Composite Materials. 2009;**43**(17):1803

[10] Mastral JF, Berrueco C, Gea M, Ceamanos J. Catalytic degradation of high density polyethylene over nanocrystalline HZSM-5 zeolite. Polymer Degradation and Stability. 2006;**91**:3330

[11] Liu P, Zao M. Hallosite nanotubes/polystyrene nanocomposites via in situ bulk polymerization. Journal of Thermal Analysis and Calorimetry. 2008;**94**:426

[12] Yarahmadi NY, Jakubowicz I, Hjertberg T. Development of poly(vinyl chloride) montmoryllonite nanocomposites. Polymer Degradation and Stability. 2010;**95**(2):132

[13] Paradise M, Goswami T. Carbon nantubes-production and industrial application. Materials and Design. 2007;**28**:1477

[14] Kumar AP, Depan D, Tomer NS, Sing RP. Nanoscale particles for polymer degradation and stabilization-trends a. future perspective. Progress in Polymer Science. 2009;**34**:7,9

[15] Samyn F, Bourbigot S, Duguesne S, Delobel R. Effect of zinc borate on the thermal degradation of ammonium polyphosphate. Thermochimica Acta. 2007;**456**:134

[16] Yu D, Kleemeir M, Wu GM, Schartel B, et al. Phosphorus and silicon/MMT containing low-melting organic-inorganic glasses improve flame retardancy of epoxy/clay composites. Macromolecular Materials and Engineering. 2011;**296**:883

[17] Kandare E, Kandola BK, Price D, Nazaré S, Horrocks RA. Study of the thermal decomposition of flame retarded unsaturated polyester resins by TG analysis and Py-GC/MS. Polymer Degradation and Stability. 2008;**93**:1996

[18] Katsoulis C, Kandare E, Kandola BK. The combined effect of epoxy nanocomposites and phosphorus flame retardant additives on thermal and fire properties od fiber-reinforced composites. Journal of Fire Sciences. 2011;**29**:361

[19] Kicko-Walczak E, Rymarz G, Gajlewicz I. The new adhesive polyester with reduced flammability for the production of modern piping system. Przemysł Chemiczny. 2013;**92**:504

[20] Kicko-Walczak E, Rymarz G, Gajlewicz I. Nanocomomposites with reduced flammability: Synergy effect of conventional flame retardant and nanoparticles in influencing the level of flame retardancy resins. Przemysł Chemiczny. 2014;**93**:914

[21] Kicko-Walczak E. Recent advances in the use borates and modified borates as fire retardant additives in polymer applications. Polymers Research Journal. 2009;**2**:455

[22] Kicko-Walczak E, Rajkiewicz M, Rymarz G. Flame retardant epoxy resins and epoxy- glass reinforced laminates, P-393205 (2010) and international procedure PCT/PL 2011/000129

[23] Kicko-Walczak E, Rymarz G, Cichy B, Stechman M. Polyester composition of reduced flammability, P-399236 (2012) and international procedure PCT/PL2012/000055

[24] Kicko-Walczak E, Rymarz G, Gajlewicz I. Modification of polymer matrices with nano-carbon additives. The way to products with high ecological and construction properties. Przemysl Chemiczny. 2017;**96**:1381

[25] EN ISO 4589-2:1999/A1:2006. Plastics—Determination of burning behaviour by oxygen index—Part 2. Ambient-temperature test

[26] ISO 5660-1:2002. Reaction-to fire tests-Heat release, smoke production and mass loss rate. Part 1. Heat release rate (cone calorimeter method)

Chapter 4

Polyester Usage in Manufacturing of Electrical and Mechanical Products and Assemblies

Ahmad Nawaz, Bilal Islam, M. Sadiq Khattak,
Liaquat Ali, Umar Saleem, Azmat Ullah,
M. Zafar Ijaz and Weiguo Mao

Additional information is available at the end of the chapter

http://dx.doi.org/10.5772/intechopen.74368

Abstract

This chapter focuses on the processes in which polyester is usually used for the manufacturing of mechanical components and assemblies. Various methods of manufacturing these products are mentioned in this chapter. These methods include wet layup method, filament winding, pultrusion, vacuum bagging and autoclave curing, resin transfer molding (RTM) and vacuum-assisted resin transfer molding (VARTM). Various production levels and properties can be achieved by polyester resin using abovementioned processes. Each process has its own benefits and disadvantages, which are discussed in this chapter. Furthermore, the use of polyester in making electrical insulation is also discussed in the chapter. Advantages and disadvantages of each impregnation technique are also explained.

Keywords: mechanical components, polyester, manufacturing, electrical insulation

1. Introduction

Polyesters are used in several applications related to high-tech applications. Polyester is used as a raw material in manufacturing of several products in several applications including automotive dashboards, panels and light covers. Similarly, in aerospace applications, polyester has been used in various assemblies and parts of airplanes, space ships and rockets. Structure and body of boats and ships are also fabricated using this material. Several techniques are practiced in industries for manufacturing of abovementioned products and applications. These techniques mainly involve mechanical parts, assemblies, and electrical insulations.

2. Manufacturing of mechanical components

Resin transfer molding (RTM) and vacuum-assisted resin transfer molding (VARTM) are used to manufacture various parts related to ships and automotives [1]. Vacuum bagging and autoclave curing is another technique mostly used in aerospace industries. In addition, filament winding and pultrusion process have the capability to manufacture cylindrical shapes, pipes, and pressure vessels using polyester and fibers. However, high-voltage and frequency insulation applications are also performed using polyester. Seven techniques are known for high-voltage insulations. In the next topics, initially RTM, VARTM, vacuum bagging and autoclave curing, matched die molding, filament winding and pultrusion are discussed. Afterwards, impregnation or insulation of electrical devices is explained.

2.1. Wet layup method

Wet layup is the most commonly used technique. In this technique, polyester resin is poured over the reinforcement [2]. In addition, usually a tool is used to properly distribute polyester resin (in polyester resin initiator and accelerator are mixed). This proper distribution of polyester fills the air pockets and spaces before getting cured. Afterwards, other reinforcements are also added with calculated resin quantity. The abovementioned procedure is repeated until desired thickness of the product is achieved. Optimum amount of resin is necessary to impregnate the fiber or reinforcement. Afterwards, proper environment and conditions are provided to cure and harden polyester resin. Glass fibers are used as reinforcement in this method.

Various sequences are also used in this procedure. Traditionally, dry fibers are also placed in the mold and resin is poured upon them. However, to achieve resin uniformity, some industries altered the procedure by predipping fibers before placing in the mold. Additionally, resin is also introduced in mold by pressure casting, capillary action and vacuum infiltration which has improved this technique. This technique is cost-effective and used to produce parts at a mass level. However, several complex shaped components cannot be manufactured using this technique.

2.2. Vacuum bagging and autoclave curing

Aerospace industry predominantly uses this technique to manufacture several components related to aircraft [3–5]. In this technique, the inner surface of mold is usually covered with a nonsticking material, usually Teflon (PTFE) (**Figure 1**). Above this nonsticky surface, prepreg plies and porous cloth are positioned. In addition, to prepreg plies and porous cloth, another cloth known as bleeder cloth is also placed [4]. This entire vacuum-based bagging mechanism is placed inside an autoclave. Inside the autoclave, a combination of optimum or suitable temperature and pressure is maintained to cure the resin. This technique cures the resin uniformly in shortest possible time. Moreover, the pressure is kept higher inside the autoclave to remove the excess resin. This technique is relatively costly in comparison to other techniques due to its lower production capacity.

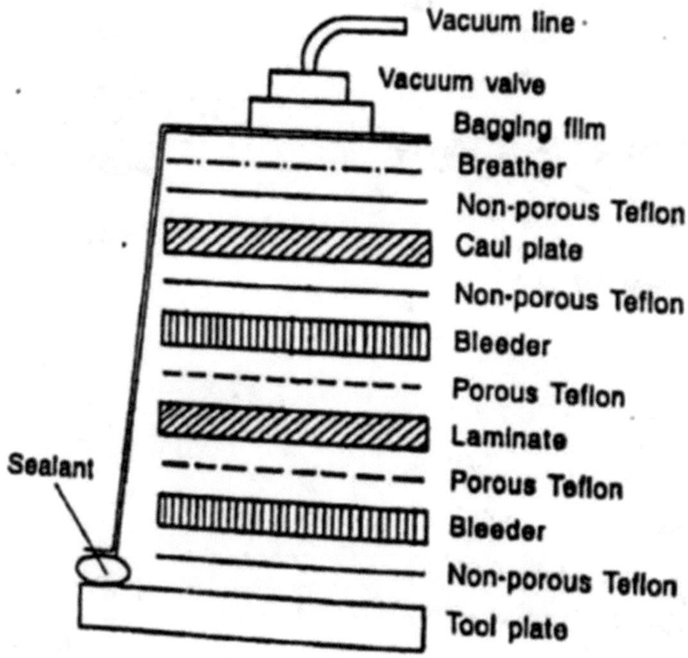

Figure 1. Schematic diagram for vacuum bagging and autoclave curing process (courtesy [4]).

2.3. Filament winding and pultrusion

Filament winding is also a composite manufacturing technique having cylindrical or tubular structures. Generally, a steel-based mandrel is manufactured in specified dimensions. Fibers are wounded upon this mandrel in a hoop, helical or polar manner. Windings are performed adjacent to one another in hoop winding (**Figure 2(a)**). However, winding can be performed at specific set of angles in the helical winding (**Figure 2(b)**) [4, 6, 7]. In contrast to helical and hoop winding, the desired product is obtained by winding between two opposite poles at specific angles (**Figure 2(c)**). Besides, there are two types of filament winding, that is, wet and dry winding. Prepreg tow is used as a winding medium in the dry winding process. On the contrary, reinforcement is properly impregnated with resin in the wet winding process. Furthermore, reinforcement impregnation is usually performed in resin bath. Later, these impregnated reinforcements are wound on the mandrel surface in a helical, hoop or polar manner.

In the filament winding process, epoxies, polyesters and silicones are usually used as polymeric resins. Water tanks, pressure vessels, automotive driving shafts, helicopter blades and rocket motor cases are generally fabricated using this technique. Furthermore, this is an economical technique which can produce these parts in varying sizes and geometries. Moreover, composites manufactured using this technique are usually void free or having less void contents.

Pultrusion is a method for production of tubular sections, rods and cylindrical type profiles in a continuous manner [2, 4]. The fundamental difference between pultrusion and filament

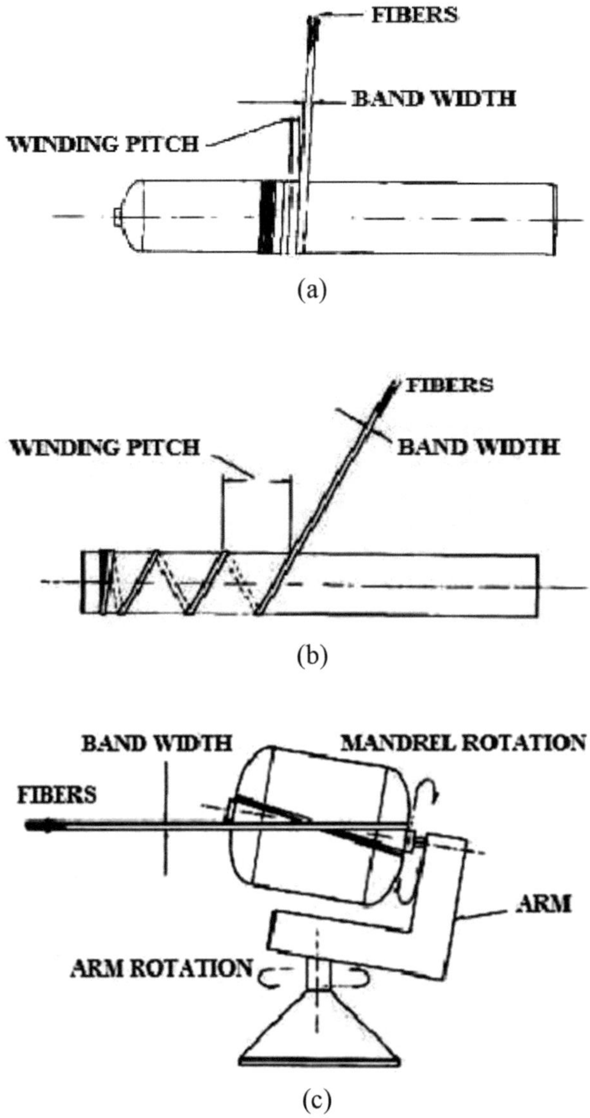

Figure 2. Schematic for filament winding patterns a) hoop, b) helical and c) polar winding (courtesy [7]).

winding is the winding pattern. In the filament winding process, the winding pattern can be hoop, helical or polar. However, pultrusion follows a longitudinal winding pattern, that is, the winding pattern is in longitudinal direction (**Figure 3**). Epoxy, vinyl ester, polyethylene and polyester-based resins are generally used in the pultrusion process. Polyester resin has been considered suitable for the pultrusion process to achieve continuous production in a short curing time. Additionally, polyester shrinks and does not stick with the mold surface after curing. Therefore, polyester resin must be preferred over other resins in pultrusion.

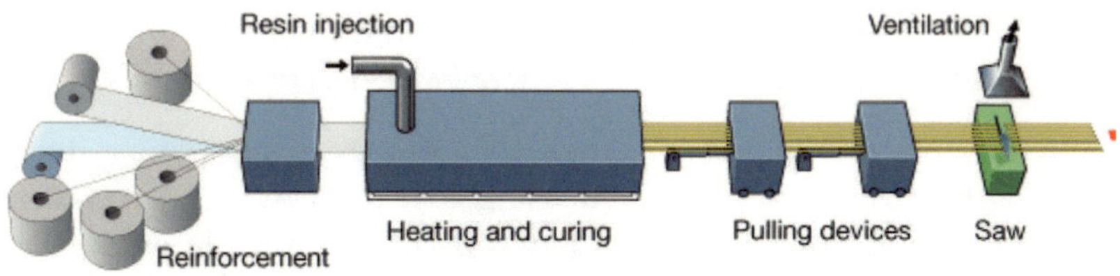

Figure 3. Schematic showing pultrusion process (courtesy Fiberline composites).

2.4. Matched die molding

To obtain uniform distribution of premixed resin, matched die molding process has been used. Reinforcement preform is placed inside the closed die mold. Afterwards, injection of premixed resin (resin in which hardener/accelerator or both are added and mixed) takes place due to externally applied pressure. Moreover, the resin enters the mold under pressure and spreads in radial direction. During the filling process, reinforcement is permeated slowly. However, operational control over the filling process is required. Improper flow leads to falling behind or racing ahead of resin. As a consequence, reinforcement is not uniformly wetted or impregnated. The final product will not be fabricated in this case. Therefore, proper understanding and prediction of flow paths are necessary for the abovementioned process. Besides, understanding the flow path, positioning of vents and injection points are also very crucial to avoid nonuniformity in flow pattern of premixed resin.

2.5. RTM and VARTM

An improved technique for producing high-quality products is RTM [8]. This technique involves insertion of low-viscosity resin into the mold. Prior to insertion in the mold, resin and hardener are mixed in the mixing chamber. Additionally, reinforcements and fabric are also placed inside the mold. Furthermore, resin is injected in mold where it impregnates the reinforcements and fabric (**Figure 4**). Afterwards, mold is heated at specific pressure and temperature to achieve the

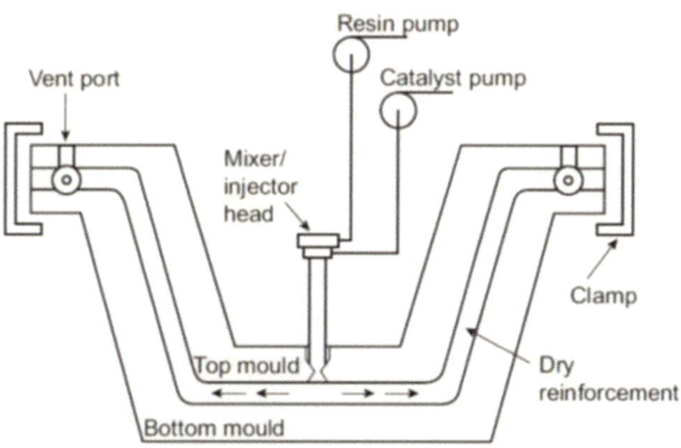

Figure 4. Resin transfer molding (RTM) process (courtesy [2]).

desired products. RTM relatively takes less curing time in comparison to other available processes. Epoxies, polyesters and vinyl esters are generally used as resins in RTM.

RTM is feasible for mass production of complex 3D shapes in comparison to prepreg processes. High-strength products are achieved using this process. No autoclave is required; therefore, RTM is a cost-effective procedure. Exposure of chemicals to workers and environment is minimized in this process. Surface finish of components made from RTM is also adequate. Training and handling costs are also reduced due to easy processing of raw material in this process. However, RTM initially involves higher tooling and investment costs. Process parameters also need to be properly controlled [1, 8, 9].

VARTM process consists of layup sealed in a vacuum bag [10]. By application of vacuum pressure, the resin is drawn inside the vacuum bag where fiber preform is placed (**Figure 5**). Resin impregnates fiber preform and starts curing. High-quality products are achieved using VARTM process due to reduction in void contents.

2.6. Electrical insulation

High-voltage power equipment is insulated by using various methods of impregnation, practiced across the globe. Selection of the process is based on the quality and production requirement. The various processes for impregnation are vacuum impregnation (VI), vacuum pressure impregnation (VPI), dip and bake impregnation or flood impregnation, trickle impregnation, B stage tapes, wet windings and full encapsulating and potting.

2.7. Dip and bake or flood impregnation

It is a very simple technique used for impregnation of electrical machines. In this technique of impregnation, the part to be impregnated is dipped inside the resin or varnish (**Figure 6**). The system is given constant heating till the varnish enters into vacant spaces of the equipment to be insulated. Heating continues till the formation of bubbles halts. This further depicts curing of polymer or evaporation of solvent. The advantage of this method is that it requires very less labor training. It is easy to implement and execute for preparing impregnated parts. A

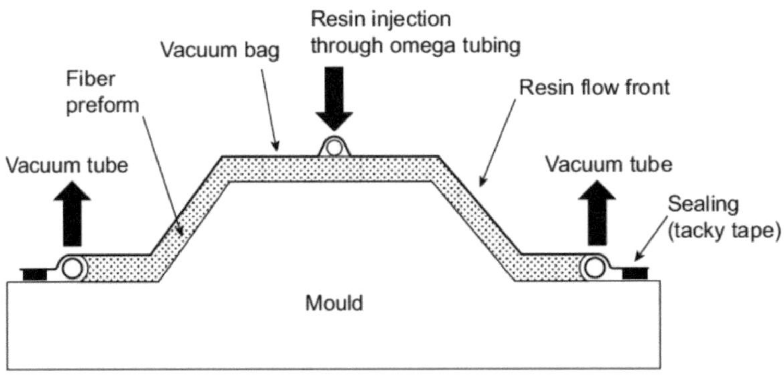

Figure 5. Vacuum-assisted resin transfer (VARTM) molding process (courtesy [2]).

Figure 6. Insulation of stator assembly with polyester resin using dip or flood impregnation (courtesy from Elantis electrical insulation pvt ltd.).

relatively less amount of investment is required in preparing manufacturing setup. It includes resin processing tank, baking oven and pneumatic system for dipping the part in varnish [11].

The disadvantage of this process is that polymer resin is dissolved in solvent. This solvent comprises 60% of volume. However, during curing process solvent evaporates and leaves behind polymer which is usually left less than 40%. As a consequence, desired insulation is not achieved. Formation of voids and bubbles is another problem in dip or flood impregnation. In other words, bubble formation percentage in insulated parts produced by dip and bake impregnation is comparatively higher than any other process. As an outcome, probability of occurrence of corona effect in voids and bubbles increases. Therefore, this process results in poor chemical and dielectric properties. Gelling of varnish or irreversible separation of varnish can be done due to the presence of acids or bases in smaller amounts. This method cannot be used in significantly high voltage insulation due to the high amount of bubble and void formation [11].

However, by using solvent-less resin, in dip and bake impregnation process, bubble percentage can be reduced to certain extent. In this process, relatively lesser material loss is observed compared to solvent-based resins. Very small amount of resin gets evaporated, that is, less than 0.5%. This makes the process also less hazardous for labor as compared to solvent-based process. Water-based solvent systems can also be used in dip impregnation, which makes it less hazardous for the labor, and there is no chance of flammability during process. In spite of using solvent-less resin, the process relatively produces insulations having higher voids and bubble contents.

2.8. Trickle impregnation

Trickle impregnation is another process performed to insulate generators, wound motors and coils. The polymer blend is poured usually with the help of nozzle on the electrical component

Figure 7. Stators insulation by trickle impregnation technique (courtesy from Elantis electrical insulation pvt ltd.).

to be insulated. Desired volume of blend is poured on the product and allowed to trickle. Turn by turn, pouring of polymer is performed by trickling (**Figure 7**). Operator usually uses paint brushes to level the polymer which is used for lamination in this specific case. In this process, windings are heated electrically to evaporate moisture contents and to reduce viscosity in certain range. In this technique, a turntable is employed to rotate the electrical components which need insulation [11]. This turntable also helps in accomplishing process quickly. In this process, void or bubble contents are reduced in comparison to dip and bake process. However, rapid changing in mixing tank makes the process less economical. Additionally, rigorous training sessions and demand for skilled labor make the process highly uneconomical. The percentage of voids in this case depends upon the process sequence (pretreatment, mixing and curing), parameters (premixing temperature, curing temperature, heating procedure and pressure during process) and the skill level of labor. The components manufactured or insulated from this technique cure more rapidly in comparison with dip and bake impregnation.

2.9. Vacuum impregnation

There is very close resemblance between dip and bake and vacuum impregnation (VI) process. Additional step in VI for proper penetration of pretreated polymer or polyester resin inside the mold is achieved by application of vacuum pressure. VI process is composed of two types of tanks [11, 12]. One is process tank, and the other is known as storage tank. A fix volume of polymer or varnish is placed inside storage tank to minimize the chances of air entrapment. Electrical components in manufacturing phases are placed inside the process tank. These electrical components are heated to eliminate the possible volatile components or wet vapors prior to resin infusion. Solvent-based resins are not recommended in this type of impregnation because when vacuum is produced in the chamber the volatile solvent starts evaporating. Vacuum is created in the process tank for a long time, ranging from 0.5 to 1 h. After that a valve which joins the storage tank with process tank is unlocked. Varnish starts entering from storage tank into process tank due to pressure difference. The valve is closed as the desired level of varnish in the tank is achieved. Vacuum effect will remain for some time, and bubbles will rise in the products. Vents are generally provided in process tank to remove the entrapped air. Products are drained and placed in an oven to bake. This process needs high initial investment. Strict maintenance activity is required at different intervals. Therefore, production is halted during the maintenance of process and storage tanks. Hence, maintenance cost of this process is relatively higher than other impregnation techniques.

Vacuum potting technique is also adopted by several industries which is a specific case of VI. In this, thermoset cross-linking polymers, that is, polyesters, vinyl esters and epoxies, are used in order to impregnate the product. Cross-linking polymers impregnate at room temperature. Here in VI, pouring in pots is done under vacuum to get bubble-free parts. Usually, ignition coils are prepared using this method. But vacuum potting is expensive compared to normal potting procedure. Furthermore, problem of bubbles and voids still remains in vacuum potting and VI process. The reason behind this bubbles or voids is due to the partial pressure which causes resin to evaporate. Evaporation of polymer also causes formation voids and bubbles in the polymeric insulation of the electric component.

2.10. Vacuum pressure impregnation

Vacuum impregnation (VI) process is used usually in tightly wounded windings. Impregnation with this process is done with highly viscous resin [11, 13]. It needs high investment and is time-consuming process. The parts manufactured by this process have high quality and are expensive due to high cycle time.

VPI shows adequate similarity with VI process. The main difference of VPI from VI is the application of higher pressure, that is, 90–100 psi after vacuum. This higher pressure is usually exerted with relatively inert gas usually nitrogen gas. In contrast, in VI process, no higher pressure is applied after vacuum (**Figure 8(a)** and **(b)**). Instead in VI, atmospheric pressure is achieved after vacuum is released. This technique shrinks bubbles and reduces their size. The process tank in VPI is designed to withstand higher pressures. Existing literature suggested that VPI is considered the best process in comparison to other known processes. VPI method is also used for making field coils (heavy coils). Reaction accelerator is also poured on the outer periphery of the insulating polymer. Due to accelerating agent, outer layer of polyester/polymer cures very quickly; this prevents polyester from spilling out of the mold or capsule. Reaction accelerator shall not be soluble in polymer resin. If reaction accelerator dissolves in polymer resin can disturb the curing and degassing process. Furthermore, accelerating chemical or agent should be reactive with resin.

2.11. B-stage tapes

B-Stage tapes are special type of insulating material which consists of partially cured resin. It is also used to insulate high voltage windings. This partially cured resin is in solid phase at low temperatures but liquefies again at room temperature. The procedure involves wounding of tapes on coils. After that it is placed in hot press machine in which it is heated and compressed. The resin is fully cured after heating up to curing temperature, which is recommended by the manufacturers. This impregnation method is used in high-voltage transformers, voltage coils, motors, and generators.

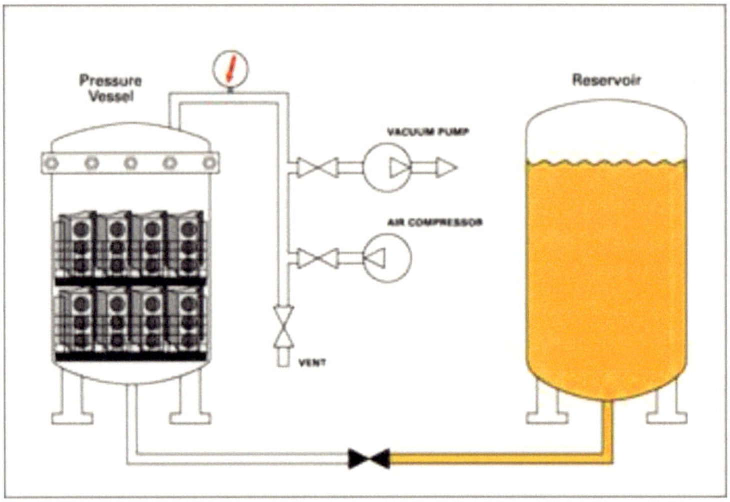

(a)

(b)

Figure 8. (a) Schematic of VPI process (Courtesy from Godfrey and Wing Inc.) (b) High-voltage assembly insertion in vacuum chamber during VPI (courtesy from Elantis electrical insulation pvt ltd.).

The problem in this type of impregnation is the high investment on hot press machine. Proper worker training is required while using this type of impregnating technique. Thirdly, this process is not labor friendly. Usual skin contact with tapes can cause skin problems. Another significant problem faced in this technique is proper cooling of tapes and taking care of their shelf life. After a certain period of time, it becomes waste and cannot be used in impregnation. So higher inventory loss is expected when compared with other processes [11, 14].

2.12. Wet windings

Wet winding is a labor-intensive and time-consuming method used in manufacturing and onsite repairs of field coils. In this impregnation process, high-viscosity pastes are applied to the windings. For curing, two types of polyester or epoxy resins are used internationally. The first type of resin is cured at a specific temperature, and the second type is cured at room temperature. It is useful in big-size coils which are difficult to prepare by other methods. Second, optimum quantity of resin is used in this wet winding process, that is, no extra resin required. Good heat transfer is observed in parts made by this method. Wet winding is a very uneconomical technique for producing smaller parts, that is, ignition devices used in automobiles, boilers, generator coils, stator and rotor insulation and so on. It requires consistent labor training. It is totally an operator-dependent process. Wet winding process is not environmental friendly [14].

2.13. Encapsulation and potting

Full encapsulating, potting and casting are similar type of processes. In all these processes, impregnation is done around a subassembly with thick and viscous resin. There is a slight difference between potting and encapsulation. In casting and encapsulation polymer (usually polyester resin premixed with hardener and accelerator is used), resin is poured in the mold in which electrical component or subassembly is also placed. Polymer cures after a certain time at room or specified temperature. Usually transformers, ignition coils and circuit boards can be made from this method. In addition, the mold or capsule does not become part of the final product (after polymer cures inside mold). On the contrary, in potting method mold or capsule usually becomes part of the final component. Heating or vacuum can be used in both these processes, but usually this type of impregnation technique is conducted at normal atmospheric conditions. The potting method is suitable for lot production of electrical components operated at higher voltages, that is, ignition coils (**Figure 9**). This process is also environmental friendly, provided hazardous solvents are not added with the resin. Potting and encapsulation are highly economical processes in comparison to the other six processes [14, 15].

Figure 9. Polyester insulated ignition coil by potting method enclosed inside polypropylene capsule (courtesy from Pecs industries Lahore (pvt) ltd.).

3. Parameters affecting quality and production

Quantity and quality of components produced from polyester using the abovementioned techniques depend upon several crucial parameters. These parameters include type of polyester resin, hardener/initiator, accelerator, inhibitor, temperature and pressure. Unsaturated polyester resins (UPE) are generally used in several applications like automotive, aerospace, boats, ships, pressure vessels and high-strength pipes. These UPE are synthesized at mass production scale by addition reaction in which glycol, that is, propylene glycol (CH_3-CH(OH)-CH_2(OH)), is reacted with unsaturated acid, that is, maleic acid (HOOC-CH=CHCOOH). Afterwards, the resultant product (alkyd polyester resins) is blended with another unsaturated monomer usually styrene or chloro-styrene. The blending ratio between styrene and unsaturated polyester resin is generally 1:2 [4]. Finally, these unsaturated polyester resins can be cured by the addition of initiator and accelerator. Methyl ethyl ketone per oxide (MEKP) and cobalt naphthionate are used as initiator/hardener and accelerator, respectively [15]. In addition, anhydrides, dimethyl and diethyl anilines are alternative materials that can serve the purpose of hardeners and accelerators (**Figure 10**).

$$ROOH + Co^{+2} \xrightarrow{Step1} RO^{*} + OH^{-} + Co^{+3}(decompostion)$$

$$ROOH + Co^{+3} \xrightarrow{Step2} ROO^{*} + H^{+} + Co^{+2}$$

$$UPE + RO^{*} \xrightarrow{Step3} UPER \xrightarrow{Step4} +styrene \xrightarrow{Step5} Polymerization$$

Figure 10. Chemical reaction for Synthesis of polyester at commercial level.

4. Conclusion

This chapter mainly discusses about various manufacturing procedures for manufacturing of composite-based mechanical and electrical parts. Prime conclusions which can be drawn from this chapter are mentioned as follows:

- Increasing or decreasing hardener and catalyst significantly affects mechanical properties of the fabricated component.
- Similarly, temperature and pressure also affect the mechanical, thermal and electrical properties of the resultant product.
- In addition, curing cycles also depend upon abovementioned parameters. Optimum ratios of these parameters can give the desired properties (mechanical, electrical, thermal and optical) and production rate.

Acknowledgements

Support is provided by Mr. Abid Iqbal and Mr. Khalid Mehmood in writing this chapter. Mr. Abid Iqbal is CEO of Pecs industries which is leading company in manufacturing of several parts for automobiles and generators. Furthermore, Pecs is vendor company of Honda, Suzuki, Fiat, Millat and Hyundai assembling plants in Pakistan. Moreover, Mr. Khalid Mehmood is CEO of Fiber craft which is leading company to manufacture composite parts and assemblies in Pakistan. However, the funding has not been provided due to certain financial crisis and higher education commission (HEC) policy.

Conflict of interest

No conflict of interest exists between the authors and their affiliations. Furthermore, no conflict of interest exists between authors and suggested potential reviewers.

Author details

Ahmad Nawaz[1]*, Bilal Islam[2], M. Sadiq Khattak[2], Liaquat Ali[1], Umar Saleem[3], Azmat Ullah[4], M. Zafar Ijaz[5] and Weiguo Mao[6]

*Address all correspondence to: nawaz.ngnr@gmail.com

1 Department of Mechanical Technology, University of Technology, Nowshera, KPK, Pakistan

2 Department of Mechanical Engineering, University of Engineering and Technology, Peshawar, KPK, Pakistan

3 Department of Mechanical Engineering, College of mechanical and electrical engineering, NUST, Rawalpindi, Pakistan

4 Department of Management Science and Engineering, Shanghai Jiaotong University, Shanghai, China

5 Department of Mechanical Engineering, Mechanical Engineering Department, Cecos University, Peshawar, Pakistan

6 Key Laboratory of Low Dimensional Materials and Application Technology, Ministry of Education, Xiangtan University, Hunan, China

References

[1] Brouwer WD, van Herpt ECFC, Labordus M. Vacuum injection moulding for large applications. Composites Part: A. 2002;**34**:551-558. DOI: 10.1016/S1359-835X(03)00060-5

[2] McIlhagger A, Archer E, McIlhagger R. Manufacturing processes for composite materials and components. In: Irving P, Soutis C, editors. Polymer Composites in Aerospace Industry, 1st ed. Woodhead Publisher; 2015. pp. 53-75. DOI: 10.1016/B978-0-85709-523-7.00003-7

[3] Hubert P, Fernlund G, Poursartip A. Autoclave processing for composites. In: Advani S, Hsiao KT, editors. Manufacturing Techniques for Polymer Matrix Composites (PMCs). Woodhead Publishing Limited; 2012. pp. 414-434. DOI: 10.1533/9780857096258.3.414

[4] Gupta MC, Gupta AP. Polymer Composites. 1st ed. New Delhi: New Age International Publishers; 2005

[5] Schlimbach J, Ogale A. Out-of-autoclave curing processes for polymer matrix composites. In: Advani S, Hsiao K-T, editors. Manufacturing techniques for polymer matrix composites (PMCs). Woodhead Publishing Limited; 2012. pp. 435-480. DOI: 10.1533/9780857096258.3.435

[6] Mack J, Schledjewski R. The filament winding process in thermoplastics. In: Advani S, Hsiao K-T, editors. Manufacturing Techniques for Polymer Matrix Composites (PMCs). Cambridge (UK): Woodhead Publishing Limited; 2012. pp. 182-208. DOI: 10.1533/9780857096258.2.182

[7] Rejab MRM, Kadirgama K, Noor MM, Sani MSM, Daud R. Modification and testing of four axes filament winding machine. International Conference on Science & Technology: Applications in Industry & Education. 2008:1505-1509

[8] Scott FN, Heath R. Resin transfer moulding for civil aircraft manufacture. SAMPE European Chapter; 1992. pp. 235-247

[9] Kendall KN, Rudd CD, Owen MJ, Middleton V. Characterisation of the resin transfer moulding process. Composites Manufacturing. 1992;**3**:235-249. DOI: 10.1016/0956-7143(92)90111-7

[10] Abraham D, McIlhagger R. Vacuum assisted resin transfer moulding for high performance carbon fibre composites. In: Proceedings of the 4th International Conference on Automated Composites, ICAC95. Nottingham; 1995. pp. 299-306

[11] Weege T. Basic Impregnation Techniques. www.scribd.com/doc/59270257/Basic-Techniques-Of-Impregnation, 1993

[12] Schwider AW, Burnham J, Buritz RS. Polyester Fibre Vacuum Impregnated Epoxy Resin for High Voltage Transformers, US Patent No3979530; 1976

[13] Sharma CP, Vadera KL, Agarwal RP. Impregnation of electrical machines with solvent less resins. Pigment & Resin Technology. 1993;**22**:10-11. DOI: 10.1108/eb043006

[14] Nawaz A. Study of the Problem of Void Formation in Electrical Lamination Parts and its Removal. MS Thesis. Peshawar: University of Engineering & Technology; 2013

[15] Islam B. Study of parameters reducing curing time for hot plate glass capsule impregnation. MS Thesis. Peshawar: University of Engineering & Technology; 2016

Chapter 5

A Case Study: Particulate-Filled Polyester Hybrid Laminated Composites

Muhammad Azeem Munawar,
Dirk Wolfram Schubert, Shahzad Maqsood Khan,
Nafisa Gull, Atif Islam,
Muhammad Atiq Ur Rehman and Monika M. Voigt

Additional information is available at the end of the chapter

http://dx.doi.org/10.5772/intechopen.73476

Abstract

The aim of this work was to develop the novel glass fiber–reinforced polyester hybrid composites (PHCs) filled with micro-sized titania (TiO_2) particles and investigate their functional, mechanical and thermal behaviors. To equip PHCs of unsaturated polyester resin (UPR) with multifunctional characteristics, TiO_2 particles (1–5 wt.%) were dispersed with high disperser homogenizer using hand lay-up process (HLUP), combined with compression molding technique (CMT). The interactions (cross linking and hydrogen bonding) between polymeric chains, styrene, silica contents of glass fiber and TiO_2 particles in PHCs were confirmed by Fourier transform infrared spectroscopy (FTIR). The mechanical and thermal properties increased brilliantly by potential utilization of TiO_2 particles. The 3 wt.% of TiO_2-imbedded PHCs showed remarkable progress in tensile strength (46 MPa) as well as tensile modules (2.9 GPa) relative to unloaded PHCs. The 5 wt.% of TiO_2-imbedded PHCs showed 61 and 64% increase in impact energy and hardness, respectively. Thermo-gravimetric analysis (TGA) showed that controlled PHC-0 had the mass loss up to 50%, which was restricted to 17% by using TiO_2 particles for PHC-5. Hence, it was inferred that micro-sized TiO_2 was encouraging filler for incremental valuation in functional, mechanical and thermal characteristics of PHCs. After finding the marvelous mechanical and thermal properties of PHCs, it is endorsed that these polyester composites can be tested for high strength and high temperature applications.

Keywords: hybrid composites, polyester, titania, functional, thermal, mechanical, cross linking, hydrogen bonding

1. Introduction

Composite material is a blend of at least two components having distinguishable interfaces at point of their junction, to make one macroscopic material. One component is matrix (transmitter), which transfers the load while, other is reinforcement (carrier) that bears the same load. Wood is a natural composite, where lignin (matrix) transmits the load and cellulose (reinforcement) bears stress [1]. Polymer matrix composites (PMCs) are polymer based, where thermoplastic or thermosetting polymers are responsible for matrices and continuous fibers, discontinuous fibers, chopped fibers or fillers (organic/inorganic) are performing as reinforcements [2]. If there are at least two different matrices and single reinforcement or vice versa, these composites are called hybrid composites. Fiber reinforced polymer composites comprise continuous or discontinuous fibers, woven or non-woven fiberics (carbon, glass or Kevlar fiber) as reinforcing phase in polymer resin matrices, are the largest manufactured category of the PMCs [3].

The durability problems related with traditional materials, the needs of higher speeds of construction and increasing functionality demands led to fiber reinforced polyester composites (FRPCs) with higher specific strength, light weight, environmental and corrosion resistant and requiring less maintenance during their service life. FRPCs are basically comprised by a fibrous reinforcement (usually glass, carbon or aramid fibers) embedded in a polyester matrix. These materials have applications in construction, marine, aerospace, automotive and sports industries [4, 5]. Owing to their high in-plane tensile properties, superior receptiveness to impact damage, poor interlaminar properties and compression molding are of foremost issues [6]. To solve these impediments, further research work is required. Particulate-imbedded FRPCs also called polyester hybrid composites (PHCs), with better properties by inclusion of inorganic filler in polyester matrix, is an imperative method to dig out material issues [7], even though several metal oxides are discussed in literature like zirconium oxide, zinc oxide, aluminum oxide, silicon oxide and cerium oxide [8]. However in recent era, titanium oxide/ titania (TiO_2) has gained much attraction as inorganic phase for the preparation of PHCs [9], due to its specific properties like photocatalytic, high refractive index and UV radiation absorption that can used to develop new functional organic-inorganic materials [10].

A broad spectrum of polymer matrices are available; polyester, polyimide, polysiloxane, vinyl ester, epoxy, polydimethylsiloxane but unsaturated polyester resin (UPR) is more frequent due to its superior mechanical properties, good process ability and low cost [11]. To dissolve polyester, styrene is mostly used in industry. Styrene containing polyester has relatively low viscosity and can wet the glass fibers properly. The properties of glass fiber like, high electrical and corrosion resistance with good dimensional stability are responsible for its use as reinforcement in FRPCs [12]. Additionally, styrene can crosslink with –C=C– bond of different polyester chains resulting in improved thermal and mechanical properties [13].

The association of inorganic fillers in FRPCs contributes to reduce the void contents by lowering polymerization shrinkage resulting higher mechanical and thermal properties and fillers also increase the viscosity of the matrices leading good handling during fabrication processes. Ferreira et al. studied thermal stability of an aluminized E-glass fiber/unsaturated polyester composites compared to un-metallized E-glass/polyester composites and unreinforced polyester. The residue content of the aluminized E-glass fiber reinforced composites was greater than that of un-metallized E-glass fiber composites by nearly 26 and 658% than the unreinforced polyester [14].

The main focuses of this chapter, were the fabrication of polyester hybrid composites (PHCs) using unsaturated polyester resin (UPR) with woven E-glass fiber sheets and micro-sized TiO_2 particles by hand lay-up process (HLUP) and compression molding technique (CMT), and investigating their functional, mechanical and thermal properties for high strength and high temperature applications.

2. Fabrication of polyester hybrid composites

The unsaturated polyester resin (pre-polymer of maleic anhydride and propandiol) with average molecular weight (Mw = 1500 g/mol) having of 40 ± 2% styrene content and inhibitor hydroquinone (150–200 ppm) was obtained from local commercial market. Cobalt-naphthenate (CN) as an accelerator and methyl-ethyl-ketone-peroxide (MEKP) as curing agent was acquired from Fluka Co. E-glass fiber roving strand mat used was acquired from Toray, Japan. Micro-TiO_2 particles (melting point ~1843°C and 4.23 g/cm^3 density) were purchased from Sigma Aldrich. Then micro-sized TiO_2 in predetermined weight concentrations (1–5 wt.%) were taken and dispersed in the UPR by a high speed mixer, run with 2000 rpm at 40°C for 2 h. The dispersions were cooled to room temperature following by the addition of catalyst MEKP (2 wt.%) and accelerator CN (1.5 wt.%) and then mechanical stirring for 2 min for uniform mixing of curing agent and accelerator. Unidirectional (1D) PHCs laminates were fabricated by a layer-by-layer impregnation using HLUP. Four individual sheets of cross plied woven roving E-glass fiber mat with dimensions (180 × 180 mm^2) were immersed in UPR dispersion for complete impregnation of TiO_2-imbedded UPR matrix including catalyst and accelerator, then stacked on one another by HLUP in compressed die of stainless steel (SS).

The whole assembly of PHCs laminates (along SS die) was kept in a High Temperature Melt Press (HTMP) under 5000 pounds (2.27 metric ton) at 75°C for 2 h for getting complete curing and uniform thickness and this technique called compression molding technique (CMT). Preparation of PHCs laminates was tricky practice so a plain geometry of SS was implemented for sampling. After molding in HTMP, the whole assembly of PHCs (along SS die) was cooled to room temperature at same pressure for avoiding thermal residual stresses. All PHCs were prepared with same number of fiber sheets with changing TiO_2 content from 1 to 5 wt.% and cut for different specimens using respective ASTM standards. **Table 1** shows the codes and compositions of PHCs and one thing should be noted here that we loaded TiO_2 content, MEKP content and CN content in matrix content to make total 100 percent weight of specimens. Schematic illustration of experimental set-up for fabrication of composites using hand lay-up process (HLUP) and compression molding technique (CMT) has shown in **Figure 1**.

2.1. Proposed scheme reaction for the fabrication of TiO_2-imbedded hybrid composites

During the synthesis of PHCs, H-bonding of TiO_2 and SiO_2 contents of glass fiber and cross linking of UPR is revealed with scheme reaction in **Figure 2** Polyester chains were cross-linked

Code names for HCs	Matrix content imbedded with TiO_2 (wt.%)	Fiber content (wt.%)	TiO_2 content in matrix (wt.%)
PHC-0	50	50	0
PHC-1	50	50	1
PHC-2	50	50	2
PHC-3	50	50	3
PHC-4	50	50	4
PHC-5	50	50	5

Table 1. Code names for PHCs specimens along with their matrix, fiber and titania compositions.

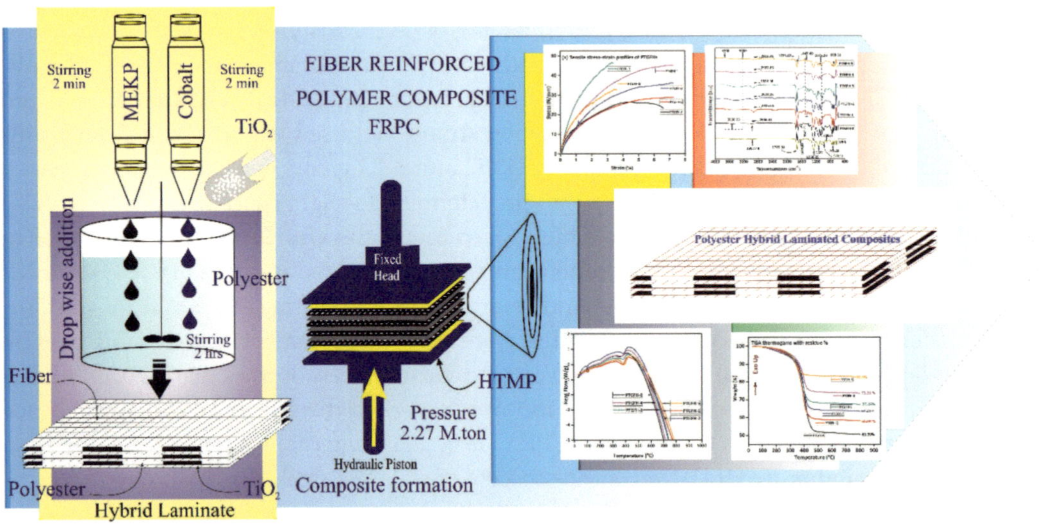

Figure 1. Schematic diagram of experimental set-up for fabrication of composites using hand lay-up process (HLUP) and compression molding technique (CMT).

with –C=C– bonds of styrene in the presence of cobalt naphthenate accelerator and MEKP catalyst. The three dimensional network of PHCs were fabricated resulting from cross linking (physical and chemical) between polyester chains, styrene, TiO_2 and SiO_2 content of glass fiber.

3. Investigation of functional, mechanical and thermal properties of PHCs

3.1. Functional testing

Spectra of PHCs were noted by Fourier transform infrared spectroscopy (FTIR) (IR Prestige-21 Shimadzu), using the attenuated total reflectance (ATR) accessary. The air background was taken before each test sample. The wavenumber range was used from 4000 to 500 cm^{-1} at resolution of 4 cm^{-1} and 100 scans per spectrum.

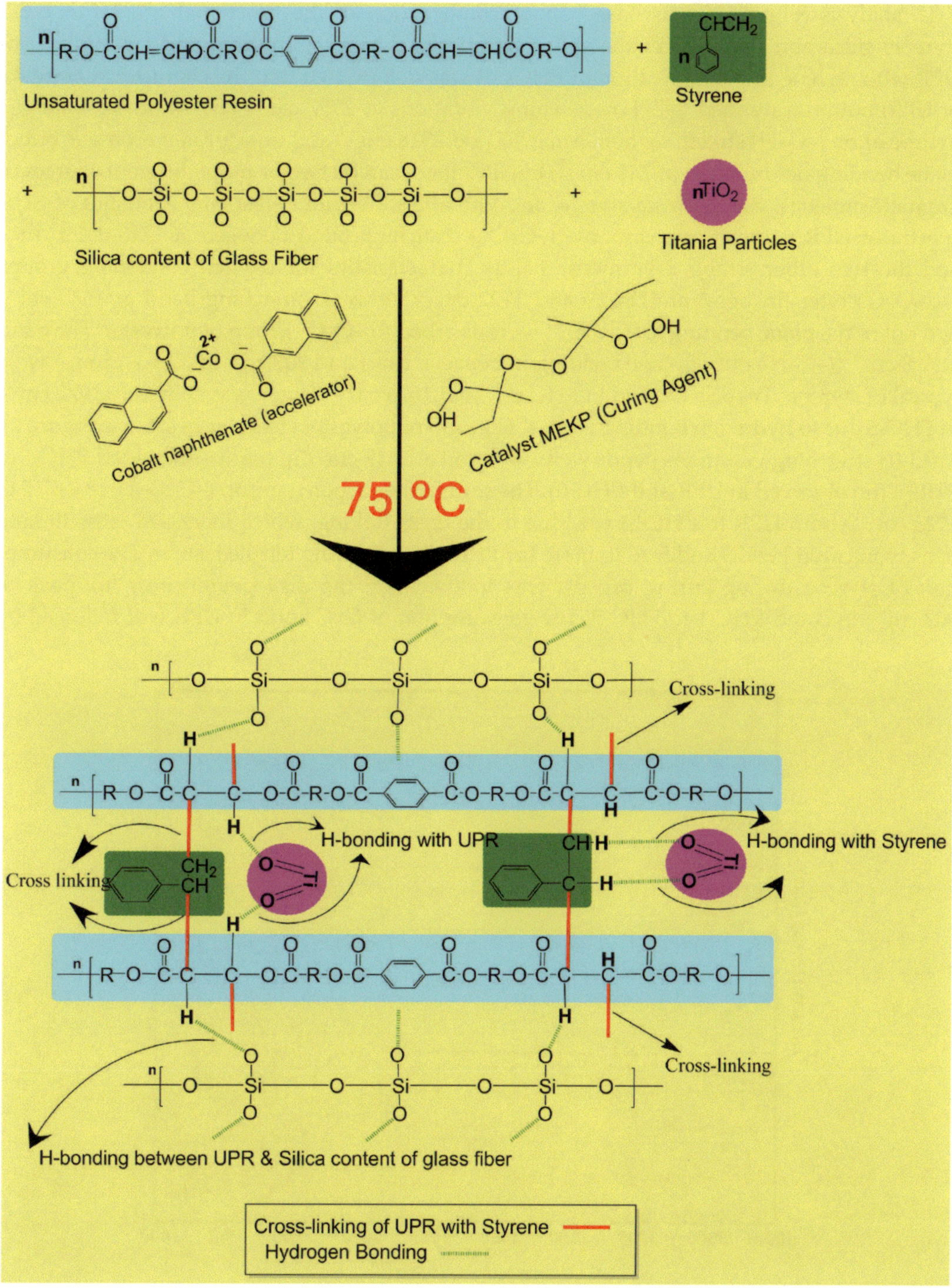

Figure 2. Proposed scheme reaction for the fabrication of TiO_2-imbedded hybrid composites in which green dotted lines (⦙⦙⦙⦙⦙) show hydrogen bonding between TiO_2 particles, unsaturated polyester resin, styrene content of UPR, silica content of glass fiber and red solid lines (—) show cross linking of unsaturated polyester resin.

FTIR analysis (**Figure 3**) was based about confirmation of the functional groups of UPR, styrene, silica and titania with the shifting of peaks during H-bonding and crosslinking of UPR with styrene in the fabrication of PHCs. It was determined that the characteristics peaks of UPR: both unsaturated (–C=H) stretching vibrations at 2957 cm^{-1} and saturated aliphatic hydrocarbon (–C–H) stretching between 2957 and 2918 cm^{-1} and unsaturated aromatic out of plane bending deformation at 741 cm^{-1}. In UPR, there can be two or more different sources of aromatic moieties: usually from styrene and Phthalic anhydride. Most characteristics of FTIR spectra of UPR is the strong carbonyl (–C=O–) stretching band between at 1715–1723 cm^{-1} and the two other strong asymmetric bands characteristics for oxygen containing groups (–C–O–C– ester linkages) at 1260.29 and 1122 cm^{-1}. A weak stretching band at 1635 cm^{-1} and out of the plane bending at 978 cm^{-1} were ascribed to –C=C– group of polyester. The band at 698 cm^{-1} (–C–H– out of plane bending in benzene ring) and 909 cm^{-1} (–C=C– group) were typical of styrene. The –Si–O band in PHC-0 was at 1076 cm^{-1} which was shifted to 1072 cm^{-1} in PHC-5 due to hydrogen bonding with –CH group of polyester chain (as in scheme **Figure 2**). Ti–O–Ti stretching vibrations bands were observed at 1447 cm^{-1} in titania-imbedded PHC-1 to PHC-5 (no observed in UPR and PHC-0). The shift in the carbonyl group (–C=O–) from 1715 to 1723 cm^{-1} (from UPR to PHC-5) was due to the cross linking, which increased cross linking density in cured PHC-5 and lead to more brittle behavior during tensile testing. The consumption of styrene during curing process was followed by the disappearance of the peak at 909 cm^{-1} in cured PHC-1 to PHC-5. The consumption of C=C bond in UPR was followed by

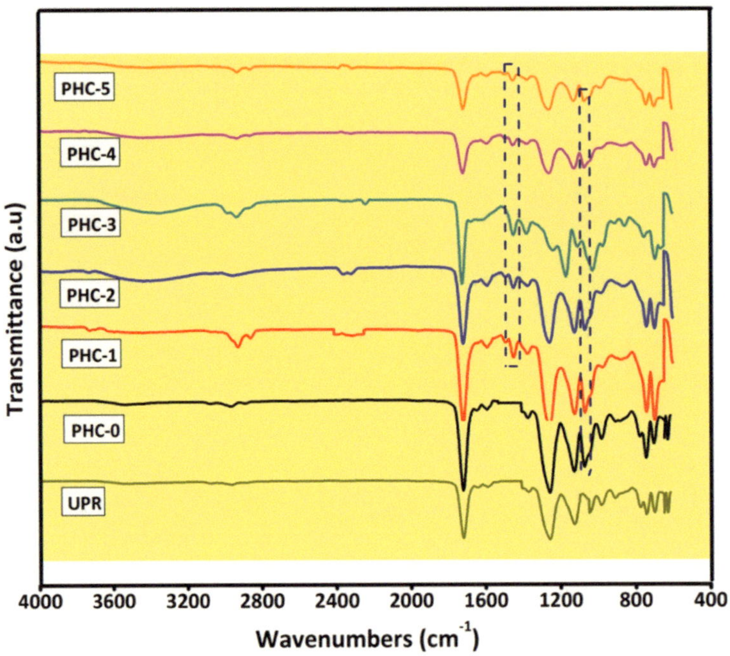

Figure 3. FTIR spectra comparison of cured UPR, controlled PHC-0 without TiO_2 particles and TiO_2-imbedded cured hybrid composites with their relative composition of TiO_2 particles; PHC-1 (1 wt.%), PHC-2 (2 wt.%), PHC-3 (3 wt.%), PHC-4 (4 wt.%) and PHC-5 (5 wt.%).

the change in the peak area at 978 cm^{-1} in PHC-0 to PHC-3 and vanished up to PHC-5. The –CH stretching band at 2957 cm^{-1} became stronger as observed in cured PHC-0 (2956 cm^{-1}) which was shifted to 2918 cm^{-1} in TiO_2-imbedded PHCs, i.e., PHC-1 to PHC-5 and the shifting of –CH band to lower wavenumber due to H-bonding with the inclusion of silica and titania loading. The stretching band at 1635 cm^{-1} (UPR) was noted at 1645 cm^{-1} in cured PHC-0 and almost vanished in PHCs having concentration of TiO_2 from 1 to 5 wt.% (PHC-1 to PHC-5). This confirms the transformation of this group to alkane through cross-linking which is credited to the contribution of –HC=CH– functional group during curing method. The all spectrum of PHCs were differed from UPR especially in the region 3700–3200 cm^{-1} (most evidently at 3550 cm^{-1}), which were the frequencies associated with –OH stretching and had been almost continuous absorption of moisture from open air [15–17].

3.2. Tensile testing

Testometric universal testing machine (UTM), Model FS100 CT UK, with load cell (100 KN) was employed to evaluate the tensile properties under static force. PHCs specimens with end tabs for the evaluation of tensile behavior were prepared following the ASTM:D 3518-0 with dimensions (200 × 25 × 2.5 mm). The UTM was controlled at room temperature conditions at a speed of 2.0 mm/min. **Figure 4(a)** exhibited tensile stress–strain profiles which showed nonlinear strains with stress on applying force. That non-linearity and asymmetrical behavior was responsible for asymmetric fiber breaking during applied stress. **Figure 4(b)** and **(c)** showed that tensile strength (TS) and tensile modulus (TM) of PHCs increased with increasing TiO_2 contents up to maximum of 3 wt.% TiO_2, which were 46 MPa (TS) and 2.94 GPa (TM) followed decreasing trend beyond that limit. It noted that 3 wt.% TiO_2 was the optimal composition for better tensile properties of TiO_2-imbedded PHCs. The improvement in the tensile properties was the outcome of creation of chemical bonds at the interface between the PHCs and TiO_2 particles [18] and optimized compositions of polyester matrix, & TiO_2 particles were key parameter of marvelous interfacial interaction [19]. When TiO_2 particles and polyester matrix are well bonded, the applied stress can be significantly shifted from the polyester matrix to TiO_2 particles and particles bear the load resulting in high tensile strength. Reduction in tensile properties beyond 3 wt.% TiO_2 was due to crack growth propagation with stress concentrations which triggered by voids in PHC-4 & PHC-5 [20].

3.3. Izod impact test and hardness

Pendulum impact tester (Model CSI-137) of Custom Scientific Instruments, was employed to evaluate the impact strength of developed composite samples in izod mode. Five individual specimens having dimension (63.5 × 12.7 × 2.5 mm) were primed for izod impact testing and notched as per ASTM: D256-10. Hardness of PHCs specimens with dimensions (25.4 × 25.4 × 2.5 mm) was investigated by Bench Rockwell hardness tester (Model NR3-DR) using ASTM:D785-08. A carbide ball indenter having spherical base of 2.5 mm diameter was infiltrate into the sample under applied force F. Applied load was 62.5 kilo pound (kp) while preload was 1000 kp while with ratio HB10. Brinell hardness value was recorded at 10 different positions on each sample. **Figure 5**, demonstrates that the resistance to impact

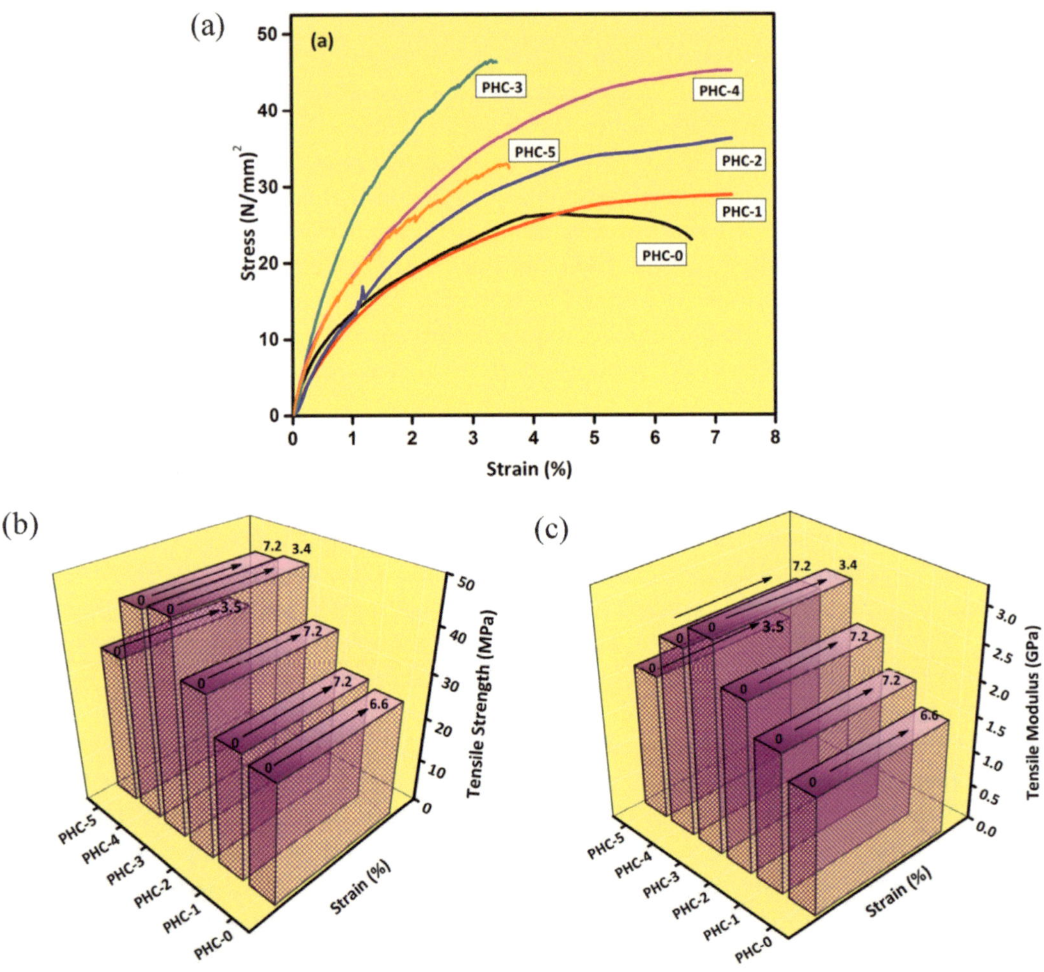

Figure 4. (a) Tensile stress-strain profiles of controlled PHC-0 and PHC-1, PHC-2, PHC-3, PHC-4, & PHC-5. (b) Show tensile strength values in mega Pascal (MPa) of PHC-0, PHC-1, PHC-2, PHC-3, PHC-4, & PHC-5 calculated from stress-strain curves. (c) Tensile modulus values in giga Pascal (GPa) of PHC-0, PHC-1, PHC-2, PHC-3, PHC-4, & PHC-5 calculated from stress-strain curves.

loading of PHCs increased with the addition of TiO_2 particles. The impact strength of PHCs was considerably depended on the particle stiffness, reinforcement strength, matrix fracture, particle-matrix interface adhesion and fiber pullout. The changing in concentration, orientation and distribution of particles in composites were also responsible for changing in physical and mechanical characteristics of resultant PHCs [21]. The level of maximum impact energy, of the order 4.73 KJ/m of PHC-5, was significantly higher with respect to unfilled PHC-0 (0.64 KJ/m). The enhancement of impact energy of PHCs was due to the strong crosslinking and hydrogen bonding (H-bonding) among TiO_2, polyester and SiO_2 content of E-glass fiber and confirmed by Ritesh Kaundal and co-workers [22]. It was noted that both tensile and flexural properties

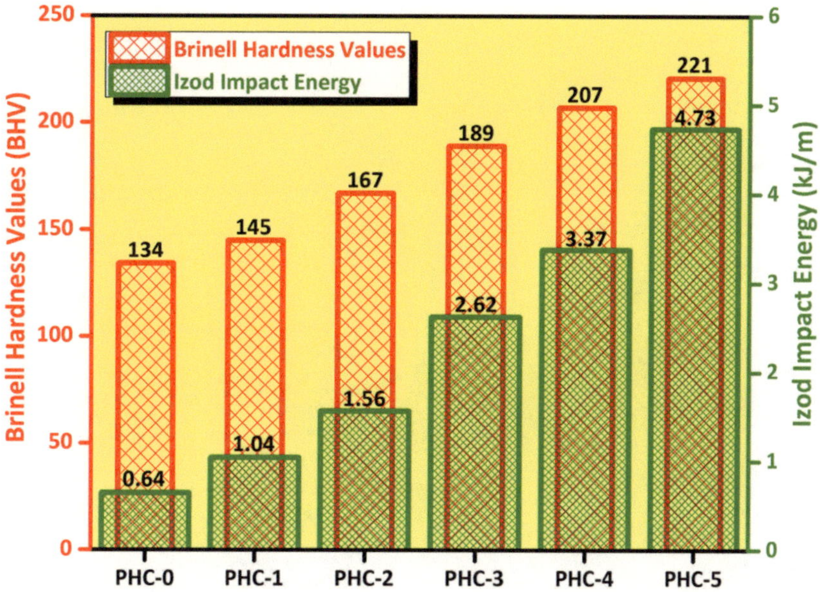

Figure 5. Izod impact strength and hardness results of PHC-0, PHC-1, PHC-2, PHC-3, PHC-4, & PHC-5.

decreased beyond 3 and 4 wt.% TiO_2 loading respectively, but impact energy values increased continuously up to 5 wt.%. It was proposed that higher concentration of TiO_2 particles produced more voids which were responsible of dissipation of impact energy more efficiently in PHC-5 [23].

Figure 5 also demonstrated that the hardness of PHCs augmented linearly with imbedding of TiO_2 particles. The PHC-5 showed 64% increase in hardness compared to PHC-0.

Improved hardness or high resistance to penetration values was due to the fact that when load was exerted on the specimens; fiber, matrix and TiO_2 particles had pressed at once and adhered with each other more effectively and interfaces transmitted pressure more efficiently. During applied compression load, the inter-particle distance got less and hardness became high. It was also considerable that matrix contents lessened on increasing the loading of TiO_2 particles in system and particles induced their intrinsic hardness to PHCs [24].

3.4. Thermal testing

Simultaneous differential scanning calorimetry/thermogravimetric analyzer (SDT) Model (Q600 of TA), was employed to conduct thermos-gravimetric analysis (TGA) of prepared PHCs. Thermal stability of each test specimen (~10 mg) was assessed at a stable temperature ramp of 10°C/min from 40 to 800°C in inert atmosphere with flow rate of 100 mL/min.

Thermal degradation pattern of controlled PHC-0 and TiO_2-imbedded PHCs depicted the weight loss in multiple steps (**Figure 6**). The weight loss from ambient to 100°C was ascribed

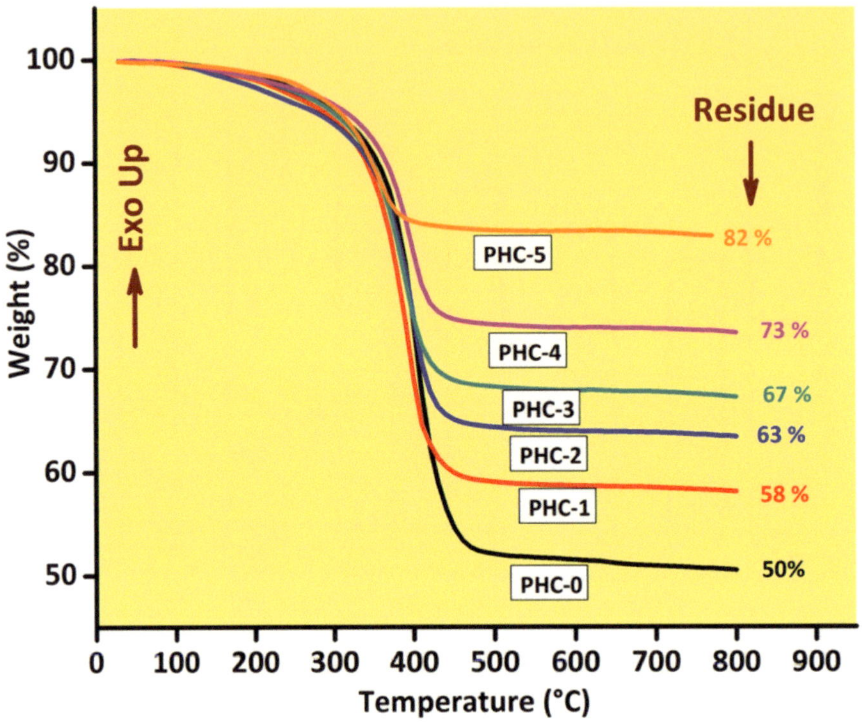

Figure 6. Thermograms of PHC-0, PHC-1, PHC-2, PHC-3, PHC-4, & PHC-5 with their relative residue contents.

to the removal of dampness and after that step dehydration took place in temperature range of 100–250°C. The onset of degradation of PHCs was started at 250°C and high degradation was linked with breaking of weak bonds. The scissoring of highly cross-linked polyester chains were observed above 350°C and transformed in un–cross-linked linear chains [25]. All straight polymer chains went to scission into tiny parts at 400°C near offset temperature [26].

The residual amount at 800°C was accredited to the left behind quantity of glass fiber and TiO_2 particles. It was observed from this result that the control PHC-0 showed the mass loss up to 50%, which was restricted to 17% for PHC-5. The PHC-5 showed 65% increase in residue contents as compared to unfilled PHC-0.

The thermal degradation data of PHCs at different steps of mass losses is given in **Table 2**. This showed that 10 and 15 wt.% loss of PHC-0 was at 352 and 365°C which improved up to 367 and 391°C in PHC-5, respectively. This data also explained that the degradation temperature was improved; overall weight loss of composite samples declined and residual amount were augmented with increasing the loading of TiO_2. The higher thermal stability was attributed to the improved interactions of polyester chains with TiO_2 and SiO_2 content of E-glass fiber and interactions decreased upon heating due to segmental movement of polyester chains [27]. As temperature slowly rose, the interface between particles and matrix started weakening which

PHCs	$T_{10\%}$[a] °C	$T_{15\%}$ °C	Residue[b] %
PHC-0	352	365	50
PHC-1	341	372	58
PHC-2	344	374	63
PHC-3	348	377	67
PHC-4	363	385	73
PHC-5	367	391	83

[a]$T_{10\%}$ and $T_{15\%}$ are the temperatures at 10% and 15% mass losses.
[b]Residue at 800°C.

Table 2. Thermal decomposition data of controlled PHC-O and TiO_2-imbedded PHCs at various percentage mass losses.

caused the motion of polymer chains radically, eventually specimens decomposed at elevated temperatures [28].

4. Conclusions

The TiO_2 embedded glass fiber reinforced polyester hybrid composites (PHCs) had fabricated by hand lay-up process (HLUP) and compression molding technique (CMT) in the presence of methyl-ethyl-ketone-peroxide (MEKP) as cross linker and cobalt naphthenate (CN) as an accelerator. It was evident that functional, mechanical and thermal characteristics of PHCs were enhanced by imbedding TiO_2 particles in unsaturated polyester resin (UPR) with E-glass fiber sheet. The formation of crosslinking of UPR with styrene molecules and hydrogen bonding between O=Ti=O particles and –Si–O content of glass fiber with UPR in PHCs were confirmed by Fourier transform infrared spectroscopy (FTIR). Tensile strength (TS) and Tensile modulus (TM)) of PHCs were improved by loading TiO_2 particles up to 3 wt.%. Impact strength and hardness improved continuously with increasing the TiO_2 loading up to 5 wt.%. TGA showed that 10 and 15 wt.% loss of PHC-0 was at 352 and 365°C, which improved up to 367 and 391°C in PHC-5, respectively. Therefore, it was concluded that TiO_2 particles are good choice as a third reinforcing material in fiber reinforced polyester hybrid composites (PHCs) as they induced the incremental variations in functional, mechanical and thermal trends. From the above conclusions of functional, mechanical and thermal behaviors, it is endorsed that these hybrid composites can be exercised for high strength and high temperature applications.

Acknowledgements

The authors are highly grateful to Higher Education Commission (HEC) Pakistan and Institute of Polymer Materials, Friedrich-Alexander-University, Erlangen-Nuremberg, 91058 Erlangen, Germany for execution of this research work.

Author details

Muhammad Azeem Munawar[1,2]*, Dirk Wolfram Schubert[2,3], Shahzad Maqsood Khan[1], Nafisa Gull[1], Atif Islam[1], Muhammad Atiq Ur Rehman[4] and Monika M. Voigt[3]

*Address all correspondence to: muhammad.munawar@fau.de

1 Department of Polymer Engineering and Technology, University of the Punjab, Lahore, Pakistan

2 Institute of Polymer Materials, Friedrich-Alexander-University, Erlangen-Nuremberg, Erlangen, Germany

3 Bavarian Polymer Institute, Fürth, Germany

4 Institute of Biopolymer Materials, Friedrich-Alexander-University, Erlangen-Nuremberg, Erlangen, Germany

References

[1] Baptista R, Mendão A, Guedes M, Marat-Mendes R. An experimental study on mechanical properties of epoxy-matrix composites containing graphite filler. Procedia Structural Integrity. 2016;**1**:74-81

[2] Agarwal G, Patnaik A, Sharma R. Thermo-mechanical properties of silicon carbide filled chopped glass fiber reinforced epoxy composites. International Journal of Advanced Structural Engineering. 2013;**5**:1-8

[3] Ribeiro MCS, Sousa SPB, Nóvoa PRO. An investigation on fire and flexural mechanical behaviors of nano and micro polyester composites filled with SiO_2 and Al_2O_3 particles. Materials Today: Proceedings. 2015;**2**:8-19

[4] Fonseca RB, de Almeida LN, Mendes GAM, Kasuya AVB, Favarão IN, de Paula MS. Effect of short glass fiber/filler particle proportion on flexural and diametral tensile strength of a novel fiber-reinforced composite. Journal of Prosthodontic Research. 2016; **60**:47-53

[5] Sayer M. Elastic properties and buckling load evaluation of ceramic particles filled glass/epoxy composites. Composites Part B: Engineering. 2014;**59**:12-20

[6] Manjunatha CM, Taylor AC, Kinloch AJ, Sprenger S. The tensile fatigue behaviour of a silica nanoparticle-modified glass fibre reinforced epoxy composite. Composites Science and Technology. 2010;**70**:193-199

[7] Nilsson F, Krückel J, Schubert DW, Chen F, Unge M, Gedde UW, et al. Simulating the effective electric conductivity of polymer composites with high aspect ratio fillers. Composites Science and Technology. 2016;**132**:16-23

[8] Munawar MA, Khan SM, Gull N, Shafiq M, Islam A, Zia S, et al. Fabrication and characterization of novel zirconia filled glass fiber reinforced polyester hybrid composites. Journal of Applied Polymer Science. 2016;**133**:43615 (1 of 9)

[9] Brandt K, Salikov V, Özcoban H, Staron P, Schreyer A, Prado LASA, et al. Novel ceramic–polymer composites synthesized by compaction of polymer-encapsulated TiO_2-nanoparticles. Composites Science and Technology. 2011;**72**:65-71

[10] Bagheri S, Mohd Hir ZA, Yousefi AT, Abdul Hamid SB. Progress on mesoporous titanium dioxide: Synthesis, modification and applications. Microporous and Mesoporous Materials. 2015;**218**:206-222

[11] Wang Y, Zhang L, Yang Y, Cai X. The investigation of flammability, thermal stability, heat resistance and mechanical properties of unsaturated polyester resin using AlPi as flame retardant. Journal of Thermal Analysis and Calorimetry. 2015;**122**:1331-1339

[12] Gull N, Khan SM, Munawar MA, Shafiq M, Anjum F, Butt MTZ, Jamil T. Synthesis and characterization of zinc oxide (ZnO) filled glass fiber reinforced polyester composites. Materials and Design. 2015;**67**:313-317

[13] Liu W, Xie T, Qiu R. Styrene-free unsaturated polyesters for hemp fibre composites. Composites Science and Technology. 2015;**120**:66-72

[14] Ferreira JM, Errajhi OAZ, Richardson MOW. Thermogravimetric analysis of aluminised E-glass fibre reinforced unsaturated polyester composites. Polymer Testing. 2006;**25**:1091-1094

[15] Cao X, Lee LJ. Control of shrinkage and residual styrene of unsaturated polyester resins cured at low temperatures: I. Effect of curing agents. Polymer. 2003;**44**:1893-1902

[16] Huang Y-J, Leu J-S. Curing of unsaturated polyester resins. Effects of temperature and initiator: 1. Low temperature reactions. Polymer. 1993;**34**:295-304

[17] Huang Y-J, Wen Y-S. Curing of unsaturated polyester resins: Effects of thickening agent. Polymer. 1994;**35**:5259-5268

[18] Chirayil CJ, Joy J, Mathew L, Koetz J, Thomas S. Nanofibril reinforced unsaturated polyester nanocomposites: Morphology, mechanical and barrier properties, viscoelastic behavior and polymer chain confinement. Industrial Crops and Products. 2014;**56**:246-254

[19] Patel VK, Dhanola A. Influence of $CaCO_3$, Al_2O_3, and TiO_2 microfillers on physico-mechanical properties of *Luffa cylindrica*/polyester composites. Engineering Science and Technology, an International Journal. 2016;**19**:676-683

[20] de Farias MA, Farina MZ, Pezzin APT, Silva DAK. Unsaturated polyester composites reinforced with fiber and powder of peach palm: Mechanical characterization and water absorption profile. Materials Science and Engineering: C. 2009;**29**:510-513

[21] Fu S-Y, Feng X-Q, Lauke B, Mai Y-W. Effects of particle size, particle/matrix interface adhesion and particle loading on mechanical properties of particulate–polymer composites. Composites Part B: Engineering. 2008;**39**:933-961

[22] Kaundal R, Patnaik A, Satapathy A. Effect of SiC particulate on short glass fiber reinforced polyester composite in erosive wear environment. Walailak Journal of Science and Technology. 2012;**9**:49-64

[23] Anu Gupta AK, Patnaik A, Biswas S. Effect of filler content and alkalization on mechanical and erosion wear behavior of CBPD filled bamboo fiber composites. Journal of Surface Engineered Materials and Advanced Technology. 2012;**2**:149-157

[24] Antunes P, Ramalho A, Carrilho E. Mechanical and wear behaviours of nano and microfilled polymeric composite: Effect of filler fraction and size. Materials & Design. 2014;**61**:50-60

[25] Isa MT, Ahmed AS, Aderemi BO, Taib RM, Mohammed-Dabo IA. Effect of fiber type and combinations on the mechanical, physical and thermal stability properties of polyester hybrid composites. Composites Part B: Engineering. 2013;**52**:217-223

[26] López F, Martín MI, Alguacil FJ, Rincón JM, Centeno T, Romero M. Thermolysis of fibreglass polyester composite and reutilisation of the glass fibre residue to obtain a glass–ceramic material. Journal of Analytical and Applied Pyrolysis. 2012;**93**:104-112

[27] Hossain MK, Hossain ME, Dewan MW, Hosur M, Jeelani S. Effects of carbon nanofibers (CNFs) on thermal and interlaminar shear responses of E-glass/polyester composites. Composites Part B: Engineering. 2013;**44**:313-320

[28] Bansal RK, Mittal J, Singh P. Thermal stability and degradation studies of polyester resins. Journal of Applied Polymer Science. 1989;**37**:1901-1908

Chapter 6

Polyester Usage for Automotive Applications

Canan Saricam and Nazan Okur

Additional information is available at the end of the chapter

http://dx.doi.org/10.5772/intechopen.74206

Abstract

Textile materials used in automobiles have a large contribution in the production of automobile carpets, seating fabrics, side, roof, floor and door panels, safety belts, tires, airbags, air filters, fuel filters, insulation materials and so on. Whereas these materials are largely manufactured by using man-made fibers, polyester (polyethylene terephthalate [PET]) is the predominant fiber used in the manufacturing of automotive textiles. This chapter deals with the use of polyester fiber in automotive applications in different forms such as knitted, woven and nonwoven textile structures and as a component of composite structures discussing the basic properties and performance aspects of the fiber.

Keywords: polyester, automotive, applications, automobile carpets, seat covers, pre-assembled interior components, tires, filters, safety equipment, engine compartment items

1. Introduction

The amount of fiber usage in a standard passenger car is sizable, reaching around 25 kg; even the safety and comfort requirement can increase this amount [1]. In the automotive industry, fibers are used in the manufacturing of textile products, which are given and described in **Table 1** [2].

In the automobiles, fibers are used in different forms of textile materials such as circular knitted, warp knitted, woven and nonwoven structures. Fibers are also used as a component in multi-layer composite structures. Textile products used in automobiles are expected to fulfill different performance requirements regarding the application area. These requirements are generally fulfilled by using man-made fibers as summarized in **Table 2** [3].

As shown in **Table 2**, polyester (mainly polyethylene terephthalate—PET—but also poly-butylene terephthalate—PBT—and polyethylene naphthalate—PEN) is the predominant fiber

Textile products used in automobiles	Description of product
Car tire cords	Fabric reinforcement for car tires
Drive belts	Fabric reinforcement for automotive drive belts
Hose	Fabric reinforcement for automotive hoses
Filters	Filter media for engine filters, air intake filters, fuel filtration
Seat belts	Narrow woven safety belt fabric
Airbags	Fabric for air bags
Automobile carpets face fabric	Tufted or needle punch face fabric for floor covering
Automobile carpet backing	Primary and secondary carpet backing
Trim	Woven, knitted and nonwoven trims for bootliners, head liners, parcel shelves and door panels
Seat cover	Woven and knitted seat covers and backing fabrics

Table 1. Textile products and their descriptions used in automobiles.

Textile products used in automobiles	Requirement	Fibers used
Seat covers	Abrasion and UV resistance, attractive design	Polyester, wool, polyamide, acrylic
Carpets	Light fastness, moldability	Polyamide, polyester, polypropylene
Seat belts	Tensile strength, abrasion and UV resistance	Polyester
Airbags	Ability to withstand high temperature inflation gases, durability to storage in compacted state over many years	Polyamide 6.6, Polyamide 4.6
Hoses, belts	Heat resistance, tensile strength, dimensional stability, adhesion to rubber and chemical resistance	Polyester, aramid
Composites (Headliner, Bootliner)	Stiffness, strength, light weight, energy absorbing and thermal stability	Glass, carbon, aramid, polyester and polyethylene

Table 2. Performance requirements of textile products and fibers used.

used in manufacturing of automotive textiles. It has a share of 42%, whereas polyamide 6.6 (PA 6.6) has a share of 26%. These fibers are preferred because of their good physical properties and high mechanical performances, being dyeable as well as inexpensive [1].This chapter deals with the use of polyester in automotive applications by discussing the basic properties and performance aspects of the fiber. It presents the recent studies about the production of textile-based automobile components made of polyester and recycled polyester fibers. In addition, recyclability of automotive components made of polyester fiber as well as the usage of polyester in the production of natural fiber composites enabling reuse of waste materials is indicated.

2. Properties and performance aspects of polyester with regard to automotive textiles

The physical and mechanical properties of PET make it widely preferred for textiles used in automobiles because of [4]:

- high resistance to temperature (short-term exposure from 210 to 240°C)
- high strength, hardness and good chemical resistance
- excellent sliding properties and good electrical insulation properties
- high resistance to abrasion and low absorption of humidity
- modulus range in tension is between 1700 and 17,000 MPa depending on modifications

Advantages and disadvantages of textile fibers commonly used in automotive applications are summarized in **Table 3** [1, 5, 6]. As stated before, polyester and polyamide are the leading fibers in automotive applications; however, the very low cost of polypropylene makes it attractive; thus, it is also included in the table. Besides, as a natural fiber, wool is also preferred even in low amounts in automotive applications.

Although PP seems to be a good choice for several automobile components with its low cost and lightweight, this fiber has several disadvantages in terms of use, such as low melting point and moderate abrasion resistance. Besides, difficulty in PP's dyeability overshadows the advantage of low cost [5].

When compared to PP, PET and PA fibers have better dyeability characteristics, temperature resistance and dimensional stability. For example, 3 M™ Thinsulate™ acoustic insulation

Fiber type	Advantages	Disadvantages
Polyester (PET)	• High abrasion resistance • High UV resistance • High stiffness • Low cost	• Low moisture absorbency • Low compression resilience • A little less in wearing resistance
Polyamide 6 (PA 6) and 6.6 (PA 6.6)	• High strain recovery • High elongation • Good thermal absorptivity • High toughness and wearing resistance	• Moderate UV resistance • High energy consumption for fiber production
Polypropylene (PP)	• Low density • Very low cost • Low energy consumption for fiber production • Excellent resistance to chemicals	• Low melting point • Moderate abrasion resistance • Low moisture absorbency • Low heat resistance
Wool	• Good thermal comfort • High resilience	• High cost • Low UV resistance

Table 3. Advantages and disadvantages of textile fibers commonly used in automotive applications.

material is composed of 35% of PET staple fibers and 65% PP fibers. While the material has low weight, thanks to the polypropylene fiber, the unquestionable role of PET fiber is strengthening of the material for enhancing its usability in automobiles [7].

PA has greater toughness, excellent tensile strain recovery and excellent adhesiveness when compared to PET [1]. However, PET has higher modulus, higher heat stability, higher resistance to color change, higher durability to sunlight exposure and it is less expensive than PA.

Wool fiber provides a high level of thermal comfort owing to its high moisture absorption ability. Together with its high level of resiliency, these features make wool fiber attractive and appropriate especially for seat cover fabrics. However, due to its high cost, PET has been replaced with that fiber and has become the predominant fiber used in seat cover fabrics. Thus, wool fiber is generally used only in high-end cars [5, 6]. Despite the inappropriateness of polyester fiber for use in applications where thermal comfort has priority, its low moisture absorption property can be an advantage when dimensional stability is required under changing environmental conditions (e.g. when it is used in seat belts). Although it is not shown in **Table 3**, cellulosic fibers can also be used especially in seat covers due to their good esthetic and thermal comfort properties. However, very good abrasion resistance of PET provides durability and makes that fiber appropriate for use in seat cover fabrics [5].

In addition to mechanical performance properties, PET has good sound insulation property, which is important for carpets and pre-assembled interior components such as headliners, bootliners, parcel shelves and door panels because of the requirements for increased driving and traveling comfort. PET fiber has a potential for use to increase sound transmission loss within a wide range of frequency (i.e. 100–500 Hz) similar to that achieved by using fiberglass [8]. Besides, nonwovens produced with a high percentage of hollow PET fibers (e.g. 45–50%) in the blends with PET fiber has demonstrated higher sound absorption rate when compared to the samples produced from 100% PET fibers [9, 10].

Moreover, due to the increasing environmental awareness to protect resources, reuse and recycle of the products have gained importance in the manufacturing of automobile components. Polyester fiber can be used in the recycled form in car upholstery, especially in the automobile carpets and seat covers [11, 12]. Polyester nonwovens can be recycled to produce new materials [13]. Moreover, polyester resin can be used as a matrix material for especially the natural fiber composites, enabling the reuse of waste materials used in the pre-assembled parts of the automobiles such as headliner, bootliners and parcel shelves [14, 15]. Thus, polyester can be considered as a select fiber for recycling and sustainability.

3. Use of polyester in automotive textiles and components

3.1. Automobile carpets

Carpets used in the automobiles are basically required to be durable against soil and abrasion with high color fastness and they are expected to insulate the noise as well as having a pleasing appearance [16]. In addition, environmental friendliness is a recent trend found in automobile carpets. In this regard, lightweight carpets for decreasing the total weight of the car and carpets from recycled materials have been developed for decreasing the environmental pollution [11].

Carpets used in automobiles basically can be categorized into two types: needle punched nonwoven carpets and tufted carpets. Fifty-five percent of the interior carpets are composed of nonwoven fabrics, whereas 45% are tufted carpets, depending on the car-manufacturing region or country [16, 17]. In both types of carpets, PET is mainly used in the facings [1]. On the other hand, needle punch facings are made from PET fibers in Europe and from PP fibers in the USA. In PET velour constructions, different fiber linear densities can be blended together to provide stability of the piles and esthetic appearance [17]. Carpets produced by using tufting method have backing surfaces for supporting the piles on the facing surface. In the backing surfaces, thermally bonded spunlaid PET and/or PP nonwovens are used [16]. For example, Colback® backings are thermally bonded spunlaid nonwovens, made from bi-component filaments with a PET core and either a PP or a PA surface. The product offers processing stability, high tear strength and uniform elongation as well as excellent thermal and dimensional stability [18]. For many years, automotive carpets have been produced by molding to the shape to match the dimensions where the carpet is used [19]. PET and/or PET blended nonwovens are appropriate to meet this requirement, since they can easily be molded into shape at relatively low temperatures and have outstanding dimensional stability [16].

Recently, the most important trend in automobile carpets is using recycled PET in order to reduce the environmental pollution and promote sustainability and efficient use of resources. The analysis of sound transmission loss of needle punch nonwoven automotive carpets made up of recycled PET revealed that comparable quality levels and specifications could be achieved using recycled PET instead of pure PET in terms of mechanical properties [11].

Made from recycled PET, Freudenberg Performance Materials from Germany produces automobile carpets with an environmentally friendly production process. In addition to using recycled fiber, they have eliminated the use of chemical binders and also obtained lighter carpets when compared to conventional automobile carpets, which promotes further environmental protection [12].

3.2. Seat covers

The required characteristics of the seat covers used in automobiles can be durability, soil resistance, UV resistance and appearance retention [17].

Automobile seats are produced in a three-layer structure. Seat cover is at the uppermost layer, a foam layer is in the middle and a scrim backing at the bottom, and these three different components are connected with adhesive layers. The seat cover at the top layer is usually made from woven, warp knitted or circular knitted fabrics, in which PET is used as a dominant fiber. Since late 1990s, PET has been used in 95% of seat covers in automobiles due to its high strength and modulus, high performance against abrasion, UV radiation and heat, outstanding anti-aging performance, good shape retention and dimensional stability and low cost [20]. PET is generally used in pure form; however, sometimes it is blended with wool. Other characteristics that make PET appropriate for seating fabric are high tear resistance, easy care property and wrinkle resistance. On the other hand, very low moisture absorption capability of PET fiber (around 0.4%) is a disadvantage in terms of thermal comfort, especially in hot weather [5, 21]. In the usage of 100% PET constructions, fabrics are also produced from waste materials blended with thermoplastic to meet price points [16].

Banex is a fabric that uses a special type of PET yarn used as a seat cover that achieves the cushioning effect of springs with the machine-made warp knitting and finishing [22].

The foam layer beneath the top layer is generally made from polyurethane (PU) and acts as an adhesive agent [23]. However, the materials composed of PET and other polymeric fibers have been recently used in order to replace the foam material [17]. Nonwoven PET fabrics made from recycled fibers and the knitted structures can be used in the cover laminate instead of PU foam. Major knitted structures preferred over PU foam are the spacer fabrics, knit, multiknit and struto. Within these knit structures, multiknit comprises two stitch layers with piles in between, whereas knit consists of a stitch layer with a pile on the top [24].

The third layer at the bottom is the backing layer, which is generally produced from either PET or PA [5, 23, 25]. In the backing layer, bolster fabrics and reinforcement nonwovens such as needle punch, hydroentangled and spunbond nonwoven fabrics are preferred [17].

3.3. Pre-assembled interior components

Pre-assembled parts are the fixtures in the car other than upholstery and carpeting, which are produced by molding in shape and covered with a fabric. Among them, there are headliners, bootliners, door panels and parcel shelves. In the production of these components, knitted, woven and nonwoven fabrics are used generally as the facing layer. On the other hand, polyester can also be used as a thermoset resin with vinyl ester in the production of natural fiber composites in automotive applications such as door panel, seat backs, headliners and dashboards. Whereas the natural fiber composites produced using vinyl esters are tougher, the orthophthalic polyester provides rigid products with low heat resistance and isophthalic polyester provides moisture resistance [14, 15].

3.3.1. Headliners

Headliners are the parts which are tightly fitted into the interior roof from the rear window to the front of the car. They are usually produced by molding and are given the shapes to house sunroofs, lamps and coat hooks [16].

The characteristics, which are expected from headliners, include esthetic properties, sound absorption, thermal insulation and cushioning. Moreover, they are required to be produced from lightweight materials. The headliners are also required to be soil resistant [17].

Headliners are produced to have at least three layers, which are esthetic-facing fabric, a foam backing and the core [17]. As the facing fabric for covering, knit tricot fabrics and woven materials are used in majority. Nonetheless, nonwoven linings can also be preferred in the headliners as the facing fabric. Whereas knitted and woven linings provide better appearance, nonwoven materials are cheaper, easy to process and they show more resistance to abrasion. The other components of headliners are selected from porous structures such as PET foam sheets and fiber-reinforced porous polymer sheets, since the sound absorbing capability and heat insulation is needed [24].

On the other hand, the construction of the materials used in the production of headliners may change in different regions. For instance, regarding the facing fabrics, it is stated that half of the facing fabrics are made up of warp-knitted tricot fabrics, which are followed with needle punched and stitch bond fabrics in Europe [17]. Contrary to this, PA and PET warp-knit tricot fabric or woven materials are selected as facing fabrics to cover the headliners [16]. Nonetheless, having the ability to be easily assembled into the roof, headliners covered with dope dyed 100% PET needle punch fabrics seem to replace conventional warp-knitted fabrics. The other reason for this selection is the cost advantage, acceptable abrasion resistance and good thermal molding characteristics of the needle punch fabrics. Moreover, the weight of the needle punch fabric can be decreased by decreasing the fiber linear density. In parallel with this weight reduction, the softness of the fabric increases as well [16]. The other fabric type which drew attention is the hydroentangled fabrics because of better durability and softness. Being made up of spunlaid and hydroentangled splittable PA/PET bicomponent fibers, Evolon® fabric has excellent strength and softness. Produced using Apex technology and used in the production of headliners, Miratec fabric can copy the fabric patterns of 3D textiles and they have high strength in both horizontal and vertical directions [17, 26].

There are also some studies in which the headliner was produced solely from polyester fiber. The headliner material was formed from two layers of polyester fibers which include the binder and non-binder fibers. Whereas the layer including 20–30% binder fiber provided the loftier part and better sound absorption properties, the layer including 40–60% binder fibers provided rigidity to the headliner material [27].

In fact, polyester fiber is suggested to be a convenient material for the recycling process because of its thermoplastic characteristics. The headliners, which are produced from PET fibers, can be ground, melted and spun into new headliners [13]. Besides, needle punch PET facing fabrics were developed with PET core as an alternative to foam type materials. Although the foam type materials control the stiffness, increase the sound absorption and provide the cushioning effect, their recycling processes are complicated. By bonding directly the facing fabric to the core, the use of foam type materials can be eliminated [17].

3.3.2. Boot (trunk) liners

The boot of the car can be considered as the extension of the cabin. The coverings such as rubber matting and a carpet joined to hardboard base is preferred as the bootliners [16].

The required properties for the bootliners can be wear resistance, abrasion resistance, durability, stiffness, lightweight and ease of cleaning. Moreover, the bootliner should not have excessive recovery since this may result in a tendency to shrink over time [16] and this property gains importance within the useful life of the product.

PP and PET are the mostly used fibers in the boot linings. Whereas PP is partially stable and because of this reason less preferred, PET is more dimensionally stable since it can be molded at high temperatures.

Because of economic concerns, nonwovens are mostly used in boot and luggage compartments. Again, in the nonwoven structures, PET is the preferred fiber for bootliners. Bootliners are usually produced from needle punch fabrics for which the facing fabric is made from a staple fiber PET or needle punch fabrics in either flat or velour construction [17].

On the other hand, the sound insulation properties of the bootliners can be improved with the underlay fabric produced from fiber batts composed of recycled fibers [17]. The fibers obtained from clothing wastes can be the blends of PET, PA, PP and acrylic or natural fibers such as wool, coconut, sisal jute and hemp [16].

The other application of polyester in the bootliners is the integration of it into the natural fiber composite materials. While natural fiber composites are preferred in automobiles since they cause a reduction in weight, energy production and cost at 10, 80 and 5%, respectively, polyester is used as a matrix material for these types of composites. Even different materials other than textile materials like sunflower can be used as the core material of these composites [28].

3.3.3. *Door panel*

The door panels are the third type of preformed structures within the car interiors. The lower section of the door panel is produced as an extension of floor covering and the upper part is upholstery fabric or vinyl [16].

Having higher modulus, good heat stability, excellent resistance to color change and high durability for sunlight degradation and being less expensive, PET is used for making door panels [17].

For the panel trim in the door including the inserts or bolster, both the underlying reinforcement fabrics and lower facing fabrics can be constructed from nonwovens. The facing fabric can be selected from flat or random velour needle punch or hydroentangled materials [17].

Whereas needled fabrics produced from PP are used in the USA, the fabrics produced from fiber spun dyed PET fiber are preferred in Europe and Japan. Since interior fabrics are subjected to UV exposure, spun dyed fibers are preferred. But, in fact, spun dyed PET fiber has lower abrasion resistance when compared with PP fiber [17].

The second component of the door trim is usually made of hydroentangled fabrics. They are joined with the facing fabric or foamed PU and usually composed of 100% PET, 75% PET, 25% viscose or 50% PET/50% PP [17].

In fact, polyester is used in the door panel constructions which have parts differing from the ones explained earlier. In a study, the vehicle door panel was patented, which is mainly made up of rigid plastic panel, a paper-backing material attached with rigid plastic panel, a polyester fiber pad providing a cushioning surface, a cloth membrane and nylon adhesive membrane placed on the cloth and finally a vinyl membrane supported by cloth fabric [29].

3.3.4. Parcel shelf

The parcel shelf is the part of the car that encloses the area between the rear seats and boot. The requirements of this part are light fastness, lightweight and sound absorption to some extent.

Usually a needled fabric made up of PA, PET or PP fibers is used in the parcel shelf [16]. Often matt is included in the parcel shelf structure to increase the sound absorption.

3.4. Tires

Tires are the interface between the car and the road. The components of tires can be classified as the tread, belt package, ply, inner liner, apex, bead bundle, sidewalls and chafer [30]. Other than the rubber and steel components, textile fibers such as rayon, PA, PET or Kevlar are used in the ply component of the tire coated with rubber.

The tire ply cords are classified into three parts, which are named as bias, belted bias and radial ply cords according to their configuration. In diagonal (bias) tires, ply cords are laid at angles less than 90° to tread centerline. Belted bias tires have the belt added in the tread region. The radial tires have body ply cords, which are laid radially from bead to bead at 90° to the centerline of the tread [16]. The ply and tire ply cords transmit the braking and steering forces and withstand burst loads [30]. The requirements of the ply cords are thus tenacity, flexibility, shrinkage at high temperature, heat resistance, wear and abrasion resistance [16, 31]. Being used on the rougher roads, and requiring lower wearing resistance, PA fiber is used in bias tires as ply cord because of its excellent toughness. On the other hand, PET fiber is used in radial tires as ply cord because of having higher modulus and reducing the flat spotting [1]. PET is strong and stiff, and it provides excellent dimensional stability [16]. Moreover, it has high tenacity, good heat resistance, good wet resistance and low water absorption [32]. When compared to PA tire ply cords, the thermal shrinkage and flat spotting characteristics are superior. However, it lacks bonding with rubber when compared to PA [24]. The PET fiber type used in the tire cord is a multifilament fiber with high modulus and low shrinkage [1]. In radial tires, rayon is also used because of its superiority to PA in high-speed impact [33]. PET is also good at high-speed impacts but loses modulus and strength faster than viscose [33]. When these three fibers, PET, viscose and PA, are compared to each other, PA has the highest tenacity, whereas viscose has the least shrinkage at 160°C. On the other hand, heat generation is very low in rayon and it depends on driving conditions for the PET fiber [16].

In comparing PA 6, PA 6.6 and PET, Naskar et al. [34] applied cyclic compression and tension onto the cord-reinforced rubber composite specimens at different strain levels and time intervals on Goodrich compression and tension fatigue tester and found out that PET tire ply cords had excellent dimensional stability but poor fatigue resistance.

PET fibers are also developed to have distinguishing properties for being used as tire ply cords. A high modulus yarn was prepared by spinning polyethylene naphthalate (PEN) or other semi-crystalline PET polymers to an optimum crystallinity state. The resulting yarn had high tenacity, dimensional stability of less than 5% and shrinkage lower than 4% [35].

Finally, PET fibers were proposed to be used in the other parts of the automobile tires rather than tire ply cords. A limited twisted PET yarn having low polymerization degree was used in the belt breaker [32].

3.5. Filters

An average car involves many different types of filters to prevent air, fuel, oil and water from contaminants such as carburetor air filter, cabin interior filter, crank base breather filter, ABS wheel/brake filter, power steering filters, engine oil filters, fuel tank filter, transmission filters, wiper washer screen filters, air conditioning recirculation filter and diesel/soot filters [16].

3.5.1. Air filters

The basic air filters in the automobiles can be classified as the engine air filter and the cabin air filter. The major purpose of air filters is cleaning the air and preventing the impurities within the air, which is used by the engine during combustion. Thus, the air filters indirectly protect the components of the engine from wear. Cabin air filter prevents the airborne pollutants and allergens within the cabin and they improve the quality of cabin air.

The air intake filters are usually produced from wet-laid, resin-impregnated cellulose papers. The other media used in the air filters are PU foam, nonwovens from synthetic, natural fibers or both in hybrid systems. In fact, the usage of nonwovens in air filtering is advantageous because they are more durable and have higher bursting strength. Besides, it is possible to control the parameters such as thickness, porosity and fiber diameter [17]. Moreover, the nonwoven filter media can be constructed to have specific characteristics such as being flame retardant, antibacterial property and so on.

PET fiber can be used in different forms as a filtration media. Mainly made up of PET fiber, QualiFlo® is developed as a gradient filtration media which is produced to have a trilaminate structure. It has exceptional dirt/dust-holding capacity consisting of continuous filament PET web with filaments having a trilobal cross section in one outer section and fully bonded air-permeable high loft batt with a randomly dispersed blend of crimped PET fibers in the other outer section. Moreover, requiring no additional binder, StarWeb, which is made up of spunbond PET, was proposed to be used in filtration with Qualiflo®. StarWeb is constructed as a trilaminate filter medium containing PET fibers as in the case of Qualiflo®. Within this filter, a top layer was produced from trilobal PET spunlaid fabric, the middle layer was produced from 100% PET homopolymer and copolymer PET filaments and the bottom layer was produced from high-loft PET crimped fiber fabrics with isotropic fiber arrangement [36].

Although there are filtration media produced only from PET fiber as is discussed in previous examples, PET is also preferred to be used in nonwoven and hybrid nonwoven structures in the air filter media.

A high-capacity hybrid, multilayer automotive air filter was developed and patented [37]. This air filter was designed to have a fluid filter media containing porous natural fibers and a porous synthetic fiber media containing absorbent spunbond PET. In the natural fiber filter media, two cotton mesh layers with different densities were used in a way that the first cotton mesh, which was placed closer to receiving end of the air stream, has lower density than the second cotton mesh layer, which was placed closer to the filtered effluent air stream. In parallel with this, synthetic fiber filter media, the first spunbond PET fiber filter placed closer to the second cotton mesh has lower density than the second spunbond PET mesh placed closer to the filtered effluent air stream creating a gradient density.

3.5.2. Fuel filters

Placed between the fuel reservoir and the engine, the fuel filters are also used to protect the engine of the cars from dirt, water and the other contaminants.

The fuel filter media is usually made from PA. The reason for this is that, although the thermoplastic materials such are fluoropolymers, PET and ethylene-vinyl alcohol copolymers are resistant to fuels, the impact toughness especially at low temperatures is not found as satisfactory as PA [38]. Nonetheless, PET fiber can be used in different forms as a component of fuel filter media.

In fact, in a patent for a multilayer plastic fuel filter having at least three layers, the inner and outer layers are suggested to be produced from plastic material, which is made conductive using additives (PA6, PA6.6, PA11 and PA12). Embedded between them, PET was listed as one polymer which can be used with the other plastics which is not made conductive (fluoropolymer other than PVDH, PET or impact-modified PET) [38].

In a patent for developing depth media in-tank fuel filter with extruded mesh shell [39], it is stated that an in-tank fuel filter is developed to have two panels of filtration media. Each panel is comprised of an outer layer of extruded mesh, a pair of spunbond filtration media and an inner layer of meltblown filtration media in between. PET is used in both the spunbond and the meltblown filtration media.

Moreover, a diesel fuel filter is patented with a smoke suppressant for which the smoke suppressant is adsorbed onto the strip of nonwoven PET fiber, which is placed equidistantly around the round container [40].

3.6. Safety equipment

Seat belts and airbags are the two main items used as safety equipment in automobiles. A seat belt is used to fix the passengers on their seats and decrease the import shock by absorbing [1]. High tensile strength and stability under static and dynamic loadings are needed in seat belts. The narrow fabric used in seat belts has mostly the weave structures such as plain weave, twill weave and sateen weave used in single-layer or double-layer structures. Seat belts are manufactured in a needle loom where the weft is inserted through the warp sheen and a selvedge is formed. On the other hand, filament yarns made of PA or PET are used to produce seat belt webbing [24].

Although PA was used as a major fiber in seatbelts for years, PET has been replaced with PA due to some advantages [40]. Having higher-impact energy-absorbing capability, less discoloration against sunlight and better dimensional stability under changing temperature and humidity conditions, PET is preferred to PA [1, 41].

When the static and dynamic loads have been applied on the seatbelts made from multifilament PET and PA yarns and compared, it has been observed that PET is superior due to lower extensibility that prevents stretching of the belts under loading during impact situations and higher stiffness [24, 41, 42].

As one other important safety equipment in automobiles, airbags are required to have extremely low gas permeability by means of a combination of high-density weaving and a thin coating treatment to resist high temperature, to have high extensibility and to be durable for storage in a compacted state for many years [43].

Demands for airbag yarns have been increasing recently, as a result of global rise in safety requirements in automobiles. Airbags are usually made up from PA 6.6 filament yarns. Nonetheless, there are some attempts to find alternative fibers. There is a noticeable trend toward PET and PA 4.6 filament yarns for the airbag fabrics and sewing threads [44, 45].

In **Table 4**, characteristics of the fibers used in airbag yarns are given [44].

North America's airbag market in 2014 comprised around 45,000 tons of PA 6.6-based fabrics and an additional 4000 tons of PET-based alternatives. On the other hand, in Europe, the use of PA 6.6-based fabrics has stayed around 33,000–35,000 tons per year while the use of PET fabrics has increased. It is forecasted that around 12,000–13,000 tons of PET will be employed in European airbags by 2020 [46].

Textile materials are used in engine compartment items such as driving belts (V-belt, synchronous belts and serpentine belt), hoses (brake and clutch hoses) and lines (power steering lines and bonnet lines).

In automobiles, the mechanical parts of the engines are driven by belts. The belts used in engine compartment of automobiles are required to be resistant to fatigue, abrasion, heat, chemicals as well as have high tensile strength and good dimensional stability [16].

A typical V-belt cross-sectional scheme is shown in **Figure 1**.

In order to enhance the fatigue properties of the belts that proposed mechanical and thermal loads, cords are used for reinforcement. The major reinforcing element used in the belts is PET

Characteristic	PA 6.6	PA 6	PET
Density (kg/m^3)	1140	1140	1390
Specific heat capacity (kJ/kg°K)	1.67	1.67	1.3
Melting point (°C)	260	215	258
Softening point (°C)	220	170	220
Heat to melt (kJ/kg)	589	522	427

Table 4. Characteristics of the fibers used in airbag yarns.

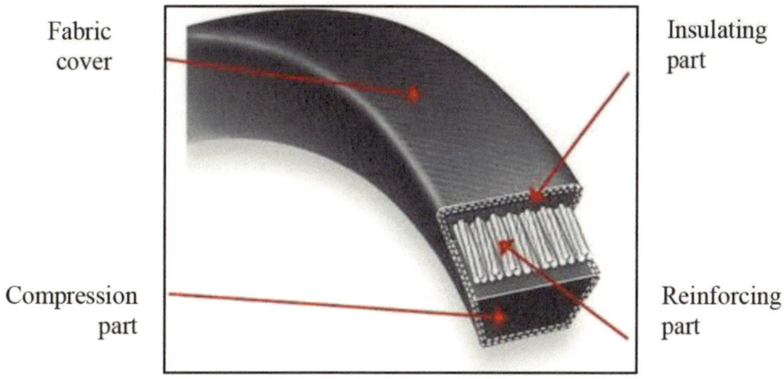

Figure 1. V-belt [47].

cord, which is composed of twisted filament yams [16]. PET fiber is applied both in the cord of V-belt and in the fabric cover of its upper part, whereas p-Aramid fiber is applied to the cords of V-belts, V-ribbed belts and metal-combined belts [1].

The bonnet line and the fabric linings with it in the engine compartment require both thermal and sound insulation functions and it usually is constructed from metal or fiber-reinforced plastic composite. PET spunbond is usually used as a nonwoven facing to cover stiffening components such as glass fiber foam or resin bonded nonwoven fabric [17].

The brake and clutch hoses are required to prevent absorption of the lubricant fluid and to resist the fluid. As an example of those kinds of hoses supplied by Fenox, the PET yarn is used as the reinforcement component that prevents fluid absorption and extends the service life of the fluid and also it improves the ability of the hose to withstand pressures as a result of increased rigidity [48].

4. Conclusion

Although attempts have been made to investigate the use of different textile fibers such as cotton, viscose, wool, acrylic, aramid and bast fibers in automobile applications, these fibers are used in a small amount by specific automobile manufacturers. The industry is still served by manmade fibers, in particular PET fiber, due to its good physical properties, enhancing mechanical performances, functionality and durability.

Polyester fiber can be used in many forms in the automotive industry, such as knitted, woven or nonwoven fabrics, especially needle punch and stitch bond fabrics used as facing materials; woven fabrics in the airbags and seat belts; nonwoven material used as a filling material for especially improving the sound insulation property in multilayer structures in the car interiors; spunbond or meltblown nonwoven structures either used solely or in combination with other layers in the filtration media; as reinforcing material in the tire cords and engine comportment items such as V-belt and hoses; and finally as a component for resin-impregnated composites either as a core material or as a matrix material.

In future, it is expected that polyester will remain as the leading fiber that will be used in the production of automotive textiles. Besides, considering the increase in natural fiber composites with the purpose of weight reduction and environmental friendliness and increasing amount of recycled material in the automobile parts, the usage of polyester resin in natural fiber composites is expected to increase. Nonetheless, the performance and functionality of materials made of polyester should be improved in parallel to the trends, probable changes in future expectations and requirements, as for the materials produced from other types of fibers and polymers.

Author details

Canan Saricam* and Nazan Okur

*Address all correspondence to: saricamc@itu.edu.tr

Istanbul Technical University, Istanbul, Turkey

References

[1] Matsuo T. Automotive applications. In: Deopura BL, Alagirusamy R, Joshi M, Gupta M, editors. Polyesters and Polyamides. Cambridge: Woodhead Publishing Limited; 2008. pp. 525-541. DOI: 10.1533/9781845694609.3.525

[2] National Composites-Network Best Practice Guide-Technical Textiles and Composite Manufacturing [Internet]. 2010. Available from: https://compositesuk.co.uk/system/files/documents/Technical%20Textiles.pdf [Accessed: 2017-10-28]

[3] Rutnagur SM. The Indian Textile Journal. 2001;**111**:5-8

[4] Štrumberger N, Gospočić A, Bartulić Č. Polymeric materials in automobiles. PROMET-Traffic & Transportation. 2005;**17**(3):149-160

[5] Fung W, Hardcastle M. Textiles in Automotive Engineering. Cambridge: Woodhead Publishing; 2001. 355 p. DOI: 10.1533/9781855738973

[6] Berber RO. Development of high performance fabrics for automotive upholstery [Thesis]. Ege University, Institute of Science and Technology; 2014

[7] 3M™ Thinsulate™ Acoustic Insulation AU2020-6 [Internet]. 2010. Available from: https://multimedia.3m.com/mws/media/879984O/3m-thinsulate-acoustic-insulation-au2020-6.pdf [Accessed: 2017-10-23]

[8] Narang PP. Material parameter selection in polyester fibre insulation for sound transmission and absorption. Applied Acoustics. 1995;**45**(4):335-358. DOI: 10.1016/0003-682X(95)00007-V

[9] Devi RP. A study on acoustic properties of polyester and hollow polyester non-woven fabrics. International Journal of Advanced Technology in Engineering and Science. 2014; **2**(01):446-459

[10] Abdelfattah A, Ghalia E, Eman R. Using nonwoven hollow fibers to improve cars interior acoustic properties. Life Science Journal. 2011;**8**(1):344-351. DOI: 10.1108/rjta-16-03-2012-b005

[11] Atakan R. Use of recycled poly(ethylene terepthalate) in needlepunched automotive carpets [Thesis]. Istanbul Technical University, Institute of Science and Technology; 2014

[12] More Ride Comfort, Lower Fuel Consumption [Internet]. 2017. Available from: https://www.freudenberg-pm.com/markets/automotive [Accessed: 2017-10-19]

[13] Davies BA. Current and past technologies for headliners including acoustics, recycling and safety. In: Proceedings of the International Congress and Exposition; 23-26 February 1998; Detroit, Michigan. DOI: 10.4271/980951

[14] Holbery J, Houston D. Natural fiber reinforced polymer composites in automotive applications. The Journal of the Minerals, Metals & Materials Society. 2006;**58**(11):80-86. DOI: 10.1007/s11837-006-0234-2

[15] Kandachar P, Brouwer R. Applications of bio-composites in industrial products. Materials Research Society. 2001;**702**. DOI: 10.1557/PROC-702-U4.1.1

[16] Mukhopadhyay SK, Partridge JF. Automotive textiles. Textile Progress. 1999;**29**(1–2):1-125. DOI: 10.1080/00405169908688876

[17] Russell SJ, Tipper MJ. Nonwovens used in automobiles. In: Shishoo R, editor. Textile Advances in the Automotive Industry. Cambridge: Woodhead Publishing; 2008. pp. 63-85. DOI: 10.1533/9781845695040

[18] Primary and Secondary Backings for Molded Car Carpets [Internet]. 2017. Available from: http://colback-nonwovens.com/en/applications/molded-car-carpets/. [Accessed: 2017-10-23]

[19] Automobile Carpets [Internet]. 2017. Available from: http://www.truck-and-car-floor-mats.com/automobile-carpets.html [Accessed: 2017-10-30]

[20] Yao. Development of the automobile seat fabric by polyester filament. In: Proceedings of the 5th International Conference on Advanced Engineering Materials and Technology; 22-23 August 2015; Guangzhou, China. DOI: 10.2991/icaemt-15.2015.112

[21] Pamuk G. The properties of fabrics used for automotive seat covers [Thesis]. Eylul University, Institute of Science and Technology; 2006

[22] Boussu F, Cochrane C, Lewandowski M. Smart textiles in automotive interiors. In: Shishoo R, editor. Textile Advances in the Automotive Industry. Cambridge: Woodhead Publishing; 2008. pp. 172-196. DOI: 10.1533/9781845695040

[23] Erth H, Gulich B. Three-dimensional textiles and nonwovens for polyurethane foam substitution in car seats. In: Shishoo R, editor. Textile Advances in the Automotive Industry. Cambridge: Woodhead Publishing; 2008. pp. 140-149. DOI: 10.1533/9781845695040

[24] Kumar RS. Textiles for Industrial Applications. New York: CRC Press; 2014. 371 p. DOI: 10.1201/b15338

[25] Fung W. Coated and Laminated Textiles. Cambridge: Woodhead Publishing; 2002. 402 p. DOI: 10.1533/9781855737518

[26] Miratec Fabric [Internet]. 2017. Available from: http://www.gznonwoven.com/news/miratec-fabric-7170571.html [Accessed: 2017-10-21]

[27] Vehicle headliner formed of polyester fibers. 2000. US Patent 6048809A

[28] Jamrichpva Z, Akova E. Mechanical testing of natural fiber composites for automotive industry. University Review. 2013;**73**:20-25

[29] Vehicle interior door panel. 1995. US Patent 5418032A

[30] Adetan DA, Oladejo KA, Fasogbon SK. Redesigning the manual automobile tyre bead breaker. Technology in Society. 2008;**30**(2):184-193. DOI: 10.1016/j.techsoc.2007.12.006

[31] Barbani D, Pierini M, Baldanzini N. FE modelling of a motorcycle tyre for full-scale crash simulations. International Journal of Crashworthiness. 2012;**17**(3):309-318. DOI: 10.1080/13588265.2012.661212

[32] Radial tire having polyester cord breaker. 1974. US Patent 3854515

[33] Saunders JH. Polyamides, fibers. In: Grayson M, editor. Encyclopedia of Textiles, Fibers, and Nonwoven Fabrics. New York: John Wiley and Sons; 1984. pp. 347-351

[34] Naskar AK, Mukherjee AK, Mukhopadhyay R. Studies on tyre cords: Degradation of polyester due to fatigue. Polymer Degradation and Stability. 2004;**83**(1):173-180. DOI: 10.1016/S0141-3910(03)00260-X

[35] High modulus polyester yarn for tire cords and composites. 2001. US Patent 5397527

[36] Baker JF. No binder required to create composite media. Filtration & Separation. 2001;**38**(3):26-28. DOI: 10.1016/S0015-1882(01)80237-8

[37] High capacity hybrid multi-layer automotive air filter. 2004. US Patent 6811588 B2

[38] Multilayer Plastic Fuel Filter having antistatic properties. 1998. US Patent 5798048

[39] Depth media in-tank fuel filter with extruded mesh shell. 1999. US Patent 5902480A

[40] Diesel fuel filters containing smoke suppressant and a method of slow releasing said suppressant. 2002. US Patent 20020170280 A1

[41] Crouch ET. Evolution of coated fabrics for automotive airbags. Journal of Coated Fabrics. 1994;**23**(3):202-220. DOI: 10.1177/152808379402300302

[42] East AJ, Polyester fibers. In: McIntyre JE, editor. Synthetic Fibers: Nylon, Polyester, Acrylic, Polyolefin. Cambridge: Woodhead Publishing; 2005. pp. 95-157

[43] Drive Safe and In Style Through Technical Textile [Internet]. 2016. Available from: http://www.fibre2fashion.com/industry-article/7070/drive-safe-and-in-style-through-technical-textile [Accessed: 2017-10-19]

[44] Mukhopadhyay SK. Technical developments and market trends of automotive airbags In: Shishoo R, editor. Textile Advances in the Automotive Industry. Cambridge: Woodhead Publishing; 2008. pp. 255-268. DOI: 10.1533/9781845695040

[45] Airbag Yarns on the Rise in Automotive Sector [Internet]. 2016. Available from: http://www.textilegence.com/en/hava-yastigi-iplikleri-otomotiv-sektorunde-yukseliste/ [Accessed: 2017-10-19]

[46] Airbag Fabrics: Consolidation Is the Key [Internet]. 2016. Available from: http://www.innovationintextiles.com/airbag-fabrics-consolidation-is-the-key/ [Accessed: 2017-10-30]

[47] Dayton V-Belt 5L540 [Internet]. 2017. Available from: [Accessed: 2017-10-23]

[48] Fenox Global Service Brake and Clutch Hoses [Internet]. 2012. Available from: [Accessed: 2017-10-24]